AF477648

20
Topics in Organometallic Chemistry

Topics in Organometallic Chemistry
Recently Published and Forthcoming Volumes

Dendrimer Catalysis

Volume Editor: Lutz H. Gade

With contributions by

D. Astruc · A. Berger · B. D. Chandler · M.-C. Daniel
L. H. Gade · J. D. Gilbertson · R. Haag · C. Hajji · J. K. Kassube
R. J. M. Klein Gebbink · G. van Koten · P. W. N. M. van Leeuwen
J. N. H. Reek · F. Ribaudo · J. Ruiz

 Springer

The series *Topics in Organometallic Chemistry* presents critical overviews of research results in organometallic chemistry. As our understanding of organometallic structure, properties and mechanisms increases, new ways are opened for the design of organometallic compounds and reactions tailored to the needs of such diverse areas as organic synthesis, medical research, biology and materials science. Thus the scope of coverage includes a broad range of topics of pure and applied organometallic chemistry, where new breakthroughs are being achieved that are of significance to a larger scientific audience.

The individual volumes of *Topics in Organometallic Chemistry* are thematic. Review articles are generally invited by the volume editors.

In references *Topics in Organometallic Chemistry* is abbreviated *Top Organomet Chem* and is cited as a journal.

Springer WWW home page: springer.com
Visit the TOMC content at springerlink.com

Library of Congress Control Number: 2006925481

ISSN 1436-6002
ISBN-10 3-540-34474-8 Springer Berlin Heidelberg New York
ISBN-13 978-3-540-34474-2 Springer Berlin Heidelberg New York
DOI 10.1007/11603788

Springer is a part of Springer Science+Business Media

springer.com

Cover design: *Design & Production* GmbH, Heidelberg
Typesetting and Production: LE-TEX Jelonek, Schmidt & Vöckler GbR, Leipzig

Printed on acid-free paper   02/3100 YL – 5 4 3 2 1 0

Topics in Organometallic Chemistry
Also Available Electronically

For all customers who have a standing order to Topics in Organometallic Chemistry, we offer the electronic version via SpringerLink free of charge. Please contact your librarian who can receive a password or free access to the full articles by registering at:

springerlink.com

If you do not have a subscription, you can still view the tables of contents of the volumes and the abstract of each article by going to the SpringerLink Homepage, clicking on "Browse by Online Libraries", then "Chemical Sciences", and finally choose Topics in Organometallic Chemistry.

You will find information about the

- Editorial Board
- Aims and Scope
- Instructions for Authors
- Sample Contribution

at springer.com using the search function.

Preface

Over the past three decades dendrimers have emerged as a novel class of macromolecules with applications in a wide range of natural sciences. They are macromolecules possessing a highly regular molecular topology based on an iterative growth sequence of branching units. Starting from a small "core" molecule, this may lead to large globular structures that possess a high degree of molecular uniformity. The iterative branching not only leads to an efficient and symmetrical growth pattern but potentially to a rapid multiplication of functional groups, making these molecules attractive platforms for chemical functionality.

Beginning with the construction of highly branched cascade oligomers in the late 1970s, the first decade of progress in the field was dominated by the development of synthetic strategies and the analytical methods for their characterization. During the past 15 years, the emphasis has shifted to the exploitation of this new type of macromolecule in a variety of applications ranging from materials science to molecular catalysis.

Since the first application of dendrimers in catalysis in the mid 1990s this field has advanced rapidly. As a consequence, catalytically active dendrimers have emerged as a class of molecular catalysts, which have substantially enriched the field of homogeneous (and to some extent heterogeneous) catalysis. A general survey of transition metal dendrimer catalysts and the way it has developed is provided by A. Berger, R. Klein Gebbink and G. van Koten from one of the laboratories that pioneered the field. This is followed by in-depth discussions of dendritic transition metal catalysis based on non-covalent catalyst-support interactions in a paper by F. Ribaudo, P. van Leeuwen and J. Reek, another group of pioneers in dendrimer catalysis. Both emphasize the possibilities of catalyst recycling using such covalent or non-covalent dendritic catalysts.

The aim of catalyst recycling and the exploitation of possible constructive interactions between catalytic sites underlie the rapidly growing field of stereoselective dendrimer catalysis. Since enantioselection is governed by small increments in the free enthalpy of activation, such transformations are particularly suited to assessing "dendrimer effects", which result from the immobilization of catalysts.

The development of dendrimer-encapsulated bimetallic nanoparticles has provided the interface with heterogeneous colloid catalysis, which is reviewed by B. Chandler and J. Gilbertson. In this field, dendrimers are being used to template and stabilize reduced colloidal particles in solution. The emphasis is placed on bimetallic particles and their application in a variety of fundamental catalytic transformations.

As cheaper and readily accessible alternatives to regular dendrimers, hyperbranched polymers are increasingly being used as catalyst platforms. Rainer Haag has been one of the leaders in this field. He and C. Hajji provide an overview of an area for which commercial applications are most likely. Finally, all of these catalysis-related topics are complemented by a review of metallodendritic exoreceptors for the redox recognition of oxo-anions and halides, written by D. Astruc. This field offers new perspectives both for catalytic transformation and the development of molecular sensors.

A decade after the initial pioneering studies, dendrimer catalysis has developed into a highly varied and complex field of research. While much of the fascination is derived from the aesthetic appeal of dendrimers (as with other applications of this class of macromolecule), dendrimer catalysts have vindicated many of the utilitarian aspects of this research. This volume provides a comprehensive overview of the current state of the art.

Heidelberg, June 2006   Lutz H. Gade

Contents

Top Organomet Chem (2006) 20: 1–38
DOI 10.1007/3418_030
© Springer-Verlag Berlin Heidelberg 2006
Published online: 23 June 2006

Transition Metal Dendrimer Catalysts

Alexsandro Berger · Robertus J. M. Klein Gebbink (✉) ·
Gerard van Koten (✉)

Faculty of Science, Organic Chemistry and Catalysis, Utrecht University, Padualaan 8,
3584 Utrecht, The Netherlands
r.j.m.kleingebbink@chem.uu.nl, g.vankoten@chem.uu.nl

Abstract The development of environmental friendly processes in catalysis has been the focus of much research over the past decade. The attachment of a metal or an organometallic moiety to a dendrimer provides nanosized catalysts with unique architectures and properties that permit catalysts to be recovered from product streams through a variety of separation technologies. This review surveys progress in the synthesis and application of recyclable metallodendrimers for homogeneous catalysis and membrane filtration technology.

Keywords Catalyst recycling · Homogeneous catalysis · Membrane filtration techniques · Metallodendrimers

1
Introduction

The incorporation of a transition metal into a dendrimer has become an exciting approach in catalysis and nanoscience. The merging of (organo)metallic species, with their useful catalytic, electrochemical, optical and magnetic properties, to dendrimers, which have their own advantages, such as functional group multiplicity, site isolation, steric environments and recyclability, provides a variety of advanced applications for these innovative organometallic macromolecules in biological, physical and chemical sciences, and especially in nanotechnologies [1–8]. A notable example is the use of dendrimers as soluble supports for homogeneous catalysts, i.e., the incorporation of main group elements, organometallic or transition metal entities into the dendritic framework; this is an approach that has undergone much development over the last decade [9–13]. The benefit of using so-called *metallodendrimers* for catalytic applications mainly comes from the potential to be able to remove the catalyst (i.e. metal) as well as to recycle it. Their nano-sized nature enables these macromolecular catalysts to be recovered from the product stream by various means, especially via nanofiltration techniques. Due to their high solubility and pseudospherical structure (which can be modulated by altering the design of the dendrimer skeleton), the catalyst loading on dendrimers can be determined exactly, thereby enabling a direct comparison to be made with unsupported molecular catalysts. These features show that, in principle, dendritic catalysts can integrate the advantages of homogeneous catalysts (mild reaction conditions, well-defined catalytic sites and high selectivities) with those of heterogeneous catalysts (recyclability, high total turnover numbers) into one chemical entity, facilitating cleaner and more selective *green* processes [14–22].

As this is an opening chapter, this review does not aim to be comprehensive, but instead aims to highlight the fundamental principles and synthesis of metallodendrimers applied in homogeneous catalysis and nanofiltration techniques, focusing on some of our own efforts and on some selected pioneering examples from the field.

Due to the magnitude of research that has been reported and the variety of dendrimers and metallodendrimers that have been developed for this purpose, throughout the chapter the reader will be directed to reviews and to the other chapters in this book for further information on selected topics.

2
Current Status

2.1
Dendrimers

Dendrimers are monodisperse, highly branched and well-defined three-dimensional macrostructures. The pseudospherical shape of a dendrimer arises from its structure, which consists of an internal region (the *core*) which is connected to repeating units constituting a radial branching pattern (Fig. 1).

Since the synthesis of "cascade" molecules by Vögtle et al. [23], several different synthetic methodologies for a variety of different kinds of dendrimers have been developed. In general, their construction relies on two approaches, described as *divergent* and *convergent* (Scheme 1) [1, 2]. The divergent strat-

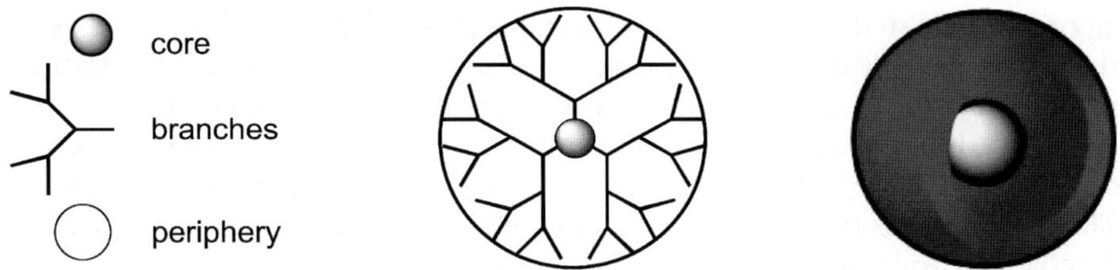

Fig. 1 Two- and three-dimensional representations of dendrimers

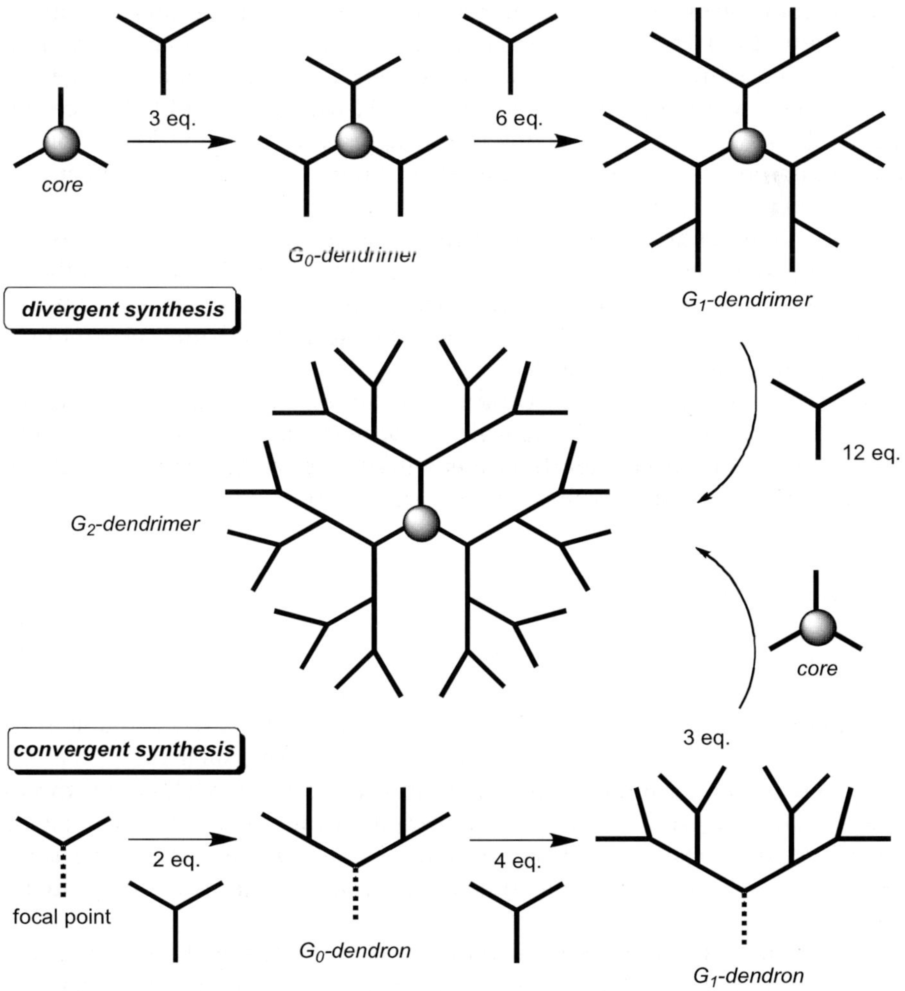

Scheme 1 Schematic representation of divergent and convergent approaches to dendrimer synthesis

egy is the most frequently applied synthetic methodology and consists of the sequential addition of repeating branching units to a core. This strategy was pioneered by Tomalia and coworkers [24], who reported the synthesis of "starburst" polyamidoamine (PAMAM) dendrimers, and by Newkome et al. [25], who reported the preparation of tree-like "arborol" dendrimers. The advantage of this approach is the easy accessibility of large quantities of dendrimers, since increasing its size (*generation*; G) also increases the molecular mass of the dendrimer. However, this approach invariably leads to the formation of dendrimers containing irreparable and irremovable branch defects resulting from a small percentage of incomplete reactions in one or more of the synthetic steps.

The convergent method was introduced by Fréchet et al. [26, 27] for the formation of poly(aryl benzyl ether) dendrimers. In this strategy, branching units are repetitively attached to the focal point of a dendron or dendritic wedge prior to its final connection to a core, which then provides the full dendritic structure. The attraction of this approach lies in the fact that a restricted number of smaller molecules are involved in the reaction steps necessary to form each generation, which statistically minimizes the formation of defects in the dendrons and facilitates the purification of the full dendrimer due to size differences in the final synthesis step. Nevertheless, low yields in the final reaction where the dendrons are attached to the core can occur due to the hindered focal groups used to form the higher dendrimer generations.

Although most of the dendrimers are synthesized by one of these two strategies, examples of metallodendrimer syntheses applying both methodologies have also been reported, which is done to improve their overall synthesis and/or purity (polydispersity) by minimizing the possible formation of defects in the dendritic framework.

2.2
Metallodendrimers

The most striking properties of dendrimers besides their appealing architecture and globular shape are their controllable size and good solubility in a large range of solvents, which make them suitable for use as soluble supports for the anchoring of (organo)metallic complexes; such complexes are applied in homogeneous catalysis. These properties are important in the fine-tuning of the retention of dendritic catalysts in membrane reactor technology. In this regard, metal-containing dendrimers have become an important research field in homogeneous catalysis and several types of metallodendrimers as well as membrane reactor strategies have been developed.

The design of metallodendrimers involves considering the position and repetition of the (catalytically active) metal site in the dendrimer framework, such as on the *periphery* (**A**) or at the *core* (**B**). Figure 2 shows schematic representations of different types of metallodendrimers.

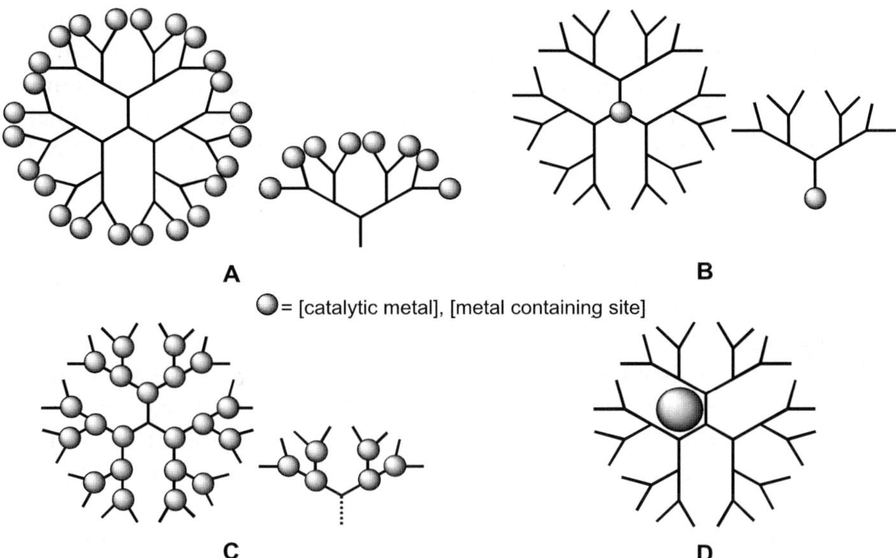

Fig. 2 Schematic representations of metallodendritic architectures according to the metal (catalyst) *location*; **A** at the periphery of a dendrimer or of a dendron; **B** at the core of a dendrimer or at the focal point of a dendron; **C** at branching points of a dendrimer or of a dendron; **D** dendrimer-encapsulated metal nanoparticles (DEMNs)

In homogeneous catalysis, the location of the metal within the dendritic framework is an important factor, since it can affect the performance of the catalyst (selectivity, activity and stability) in a "positive", "neutral" or "negative" way depending on the inherent mechanism of the reaction applied and the degree of interplay between the active sites in the nanosized species. The positioning of organometallic units at the surface of a dendrimer support is, in most cases, accessed via a divergent route, where the introduction of the metal units occurs in the last step of the formation of the metallodendrimer. In this fashion, the catalysts would in principle provide active sites that are easily accessible for substrates, which would result in reaction rates similar to unsupported homogeneous systems, provided that each can act independently. Nevertheless, these periphery-functionalized systems contain a high local concentration of multiple reaction sites, which can lead to cooperative or noncooperative effects on the catalytic sites according to the mechanism operated in these reactions (the "dendritic effect"; see Chapter by Reek et al., this volume).

Core-functionalized metallodendrimers have the advantage of creating isolated sites due to the environment of the dendritic framework. In the case of core-functionalized dendrimers, the molecular weight per catalytic site (ligand/catalyst) is higher than for periphery-functionalized dendrimers, which therefore involves higher costs from a commercial point of view. The incorporation of a metal into each generation of a dendrimer, i.e., metal centers integrated into the dendrimer skeleton at the connectors or at branching points, was actually reported in the very first example of a metallodendrimer [28, 29]. Although many examples of this kind of dendritic complexes have been reported, they have been rarely applied to catalysis [9, 10, 30, 31].

Different approaches have been developed for binding a metal or organometallic moiety to these dendrimer frameworks. In numerous coordination compounds, the dendrimer and the metal are linked through a dative metal–heteroatom bond [32], while in "organometallic" compounds the linkage between the metal and the dendritic framework is realized via σ or π metal–carbon bonding [11].

In so-called *dendrimer-encapsulated metal nanoparticles* (DEMNs), metal ions are first encapsulated either in the core or at the periphery of the dendritic structure by coordination to heteroatoms of the dendrimer. Subsequently, they are reduced to yield zero-valent metal nanoparticles encapsulated in a dendritic structure [33]. The advantage of using DEMNs as catalysts is related to several factors. Catalytic nanoparticles are commonly prepared by reducing metal salts in the presence of stabilizers (polymers, ligands and surfactants), in order to control particle size and prevent agglomeration. However, when applying such methodologies, the surfaces of the nanoparticles can be passivated by these stabilizers (sterically or electrostatically). This is a great disadvantage, since it prevents the access of substrates to the catalytic surface of the nanoparticle, reducing the catalytic efficacy. Therefore, the use of dendrimers as monodispersed templates for the formation of these nanocluster catalysts is a promising approach because it allows for kinetic control of the particle size and prevents aggregation of the nanoparticles without passivating active sites on the surface. In addition, by fine-tuning the surface functionality and porosity (or generation) of the dendrimer host, the dendrimer periphery can act as a "nanoscopic filter". The state of the art for DEMNs has been extensively reviewed [3, 4] and will be discussed further in a following chapter.

In this opening chapter, we will focus our discussion on metallodendritic catalysts that localize their catalytic functions either at the periphery or at the core of a dendritic macromolecule. These types of dendritic catalyst have been by far the most widely applied of the metallodendrimers in combination with membrane separation technologies.

3
Membrane Filtration Techniques

Anchoring (catalytically active) metal centers to a nanosized dendritic support yields a macromolecular homogeneous catalyst that can be recovered from the product stream by nanofiltration techniques. Moreover, these can, in principle, be used again or used in a continuous manner by applying them in a membrane reactor [14–20]. Several catalytic membrane reactor concepts have been developed for this purpose. In addition to various reactor designs, a multitude of metallic, inorganic and organic polymeric and enzymatic membranes have been synthesized and tested. The current state of membrane technologies in catalytic reactors has been recently reviewed by Vankelecom [22].

The recovery and recycling of catalysts by membrane filtration technology largely depend on the kind of membrane that is used and on the size and geometry of the particle to be retained. The application of organic nanofiltration membranes depends, for example, on their stability under operational conditions and on their inertness towards substrates and/or intermediates/products of the catalytic process. Factors such as pore-size distribution, charge effects, hydrophilicity or hydrophobicity and solvent polarity can greatly influence the permeability of a membrane. In general, the latter aspects are still a matter of trial and error, since for most organic membranes these data are often not available.

A crucial characteristic of a membrane is the molecular weight cutoff (MWCO) value, which is defined as the molecular weight at which 90% of the solutes are retained by the membrane. The retention factor R of solute A to be separated by the membrane is defined by the ratio of the concentration of A in the permeate to that in the retentate, as expressed in the following equation:

$$R = 1 + \left(\frac{1}{\theta}\right) \ln \left(\frac{[A_R]}{[A_R] + [A_P]}\right)$$

$[A_R]$ = concentration of retentate A
$[A_P]$ = concentration of permeate A
θ = number of exchanged reactor volumes

Membrane technology has been performed using either micro-, ultra- or nanofiltration or reverse osmosis in either batch-wise or *continuous-flow membrane reactors* (CFMR).

Passive membrane dialysis is usually applied batch-wise, since its driving-force is the difference in gradient concentration between the two solutions separated by the membrane. In this case, the solute (reactants and products: small molecules) from a hypertonic solution (the resulting solution of the catalytic reaction) permeates through the membrane to the hypotonic side (pure solvent) until equilibrium has been achieved, whereas the nanosized catalyst remains confined inside the membrane (similar to a "tea-bag"; see Fig. 3A).

On the other hand, in a CFMR process the particles are transported through the membrane due to the application of a pressure. In this way, the concentrations and residence times of reagents/substrates can be regulated,

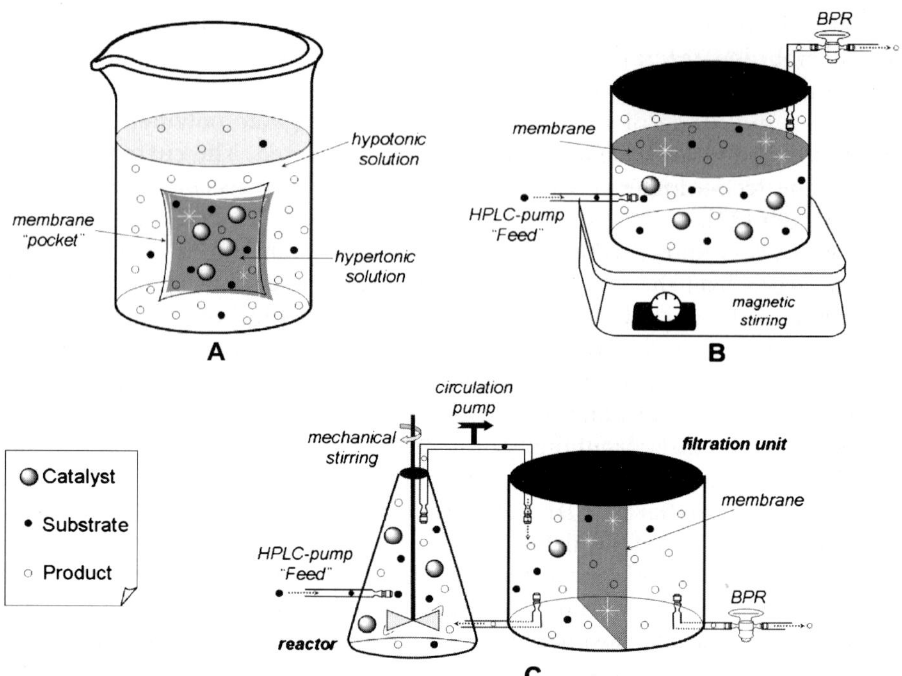

Fig. 3 Schematic representation of batch-wise passive membrane dialysis (**A**) and continuous membrane filtration: dead-end-filtration (**B**) and loop reactor (**C**)

which prevents long exposures of substrates or products to the catalytically active center and so can help to avoid the formation of side-products.

In general, two types of CFMRs are applied in homogeneous catalysis: the *dead-end-filtration* reactor (Fig. 3B) and the *loop reactor* (Fig. 3C) [19]. In the dead-end-filtration reactor the nanosized catalyst is compartmentalized in the reactor and is retained by nanofiltration membranes. Reactants are continuously pumped into the reactor, whereas small molecules (products and substrates) cross the perpendicularly positioned membrane due to the pressure exerted. Unreacted materials can be processed by adding them back into the reactor in this set-up. Concentration polarization of the catalyst near to the membrane surface can occur using this technique. In contrast, when a loop reactor is used, such behavior is prevented, since the solution is continuously circulated through the reactor and no pressure is exerted in the direction of the parallel-positioned membrane, so small particles cross the membrane laterally.

4
Metallodendrimers in Catalysis

4.1
Peripheral-Functionalized Catalysts

The benefit of applying metal-containing dendrimers in homogeneous catalysis was pioneered by the Van Koten group in collaboration with Van Leeuwen et al. [34], through the development of a synthetic methodology for the formation of zeroth- and first-generation Ni-containing carbosilane dendrimers, G_0-1 and G_1-1, respectively, which were synthesized via the divergent approach (Scheme 2). The dendrimer support comprised a so-called *carbosilane* dendrimer framework, which primarily involves a network of sigma carbon–carbon and carbon–silicon bonds.

Their advantage over other types of dendrimers is their straightforward synthesis and, most importantly, their chemical and thermal stabilities. Two distinct steps characterize their synthesis: a) an alkenylation reaction of a chlorosilane compound with an alkenyl Grignard reagent, and b) a Pt-catalyzed hydrosilylation reaction of a peripheral alkenyl moiety with an appropriate hydrosilane species. Scheme 2 shows the synthesis of catalysts G_0-1 and G_1-1 via this methodology. In this case, the carbosilane synthesis was followed by the introduction of diamino-bromo-aryl groupings as the precursor for the arylnickel catalysts at the dendrimer periphery. The nickel centers of the so-called NCN-pincer nickel complexes were introduced by multiple oxidative addition reactions with $Ni(PPh_3)_4$.

These Ni-containing dendrimers were successfully employed in the regioselective Kharasch addition of polyhalogenoalkanes to terminal $C = C$ double bonds, a reaction that was previously developed for the mononuclear NCN-pincer nickel compound. The similar product formation and reaction rates to those obtained with the monomeric catalyst suggested that each catalytic site in these dendritic hybrid catalyst acts as an independent catalytic unit. Moreover, these metallodendritic catalysts showed two other important characteristics. Firstly, metal leaching from these metallodendrimers was not observed, which is ascribed to the general stability of pincer-type organometallic complexes. Secondly, these highly soluble metallodendrimers are nanosized, which makes them amenable for homogeneous catalytic applications in nanofiltration membrane reactor set-ups. Overall, this report on carbosilane-supported catalysts triggered a new research area in the development of catalytically active metallodendrimers for homogeneous catalysis.

Based on these preliminary results, a small library of NCN-pincer nickel-containing metallodendrimers was prepared by Van Koten et al. in order to investigate the factors that can affect the catalyst performance and their applicability in nanofiltration membrane reactors [35, 36]. The strategy in this

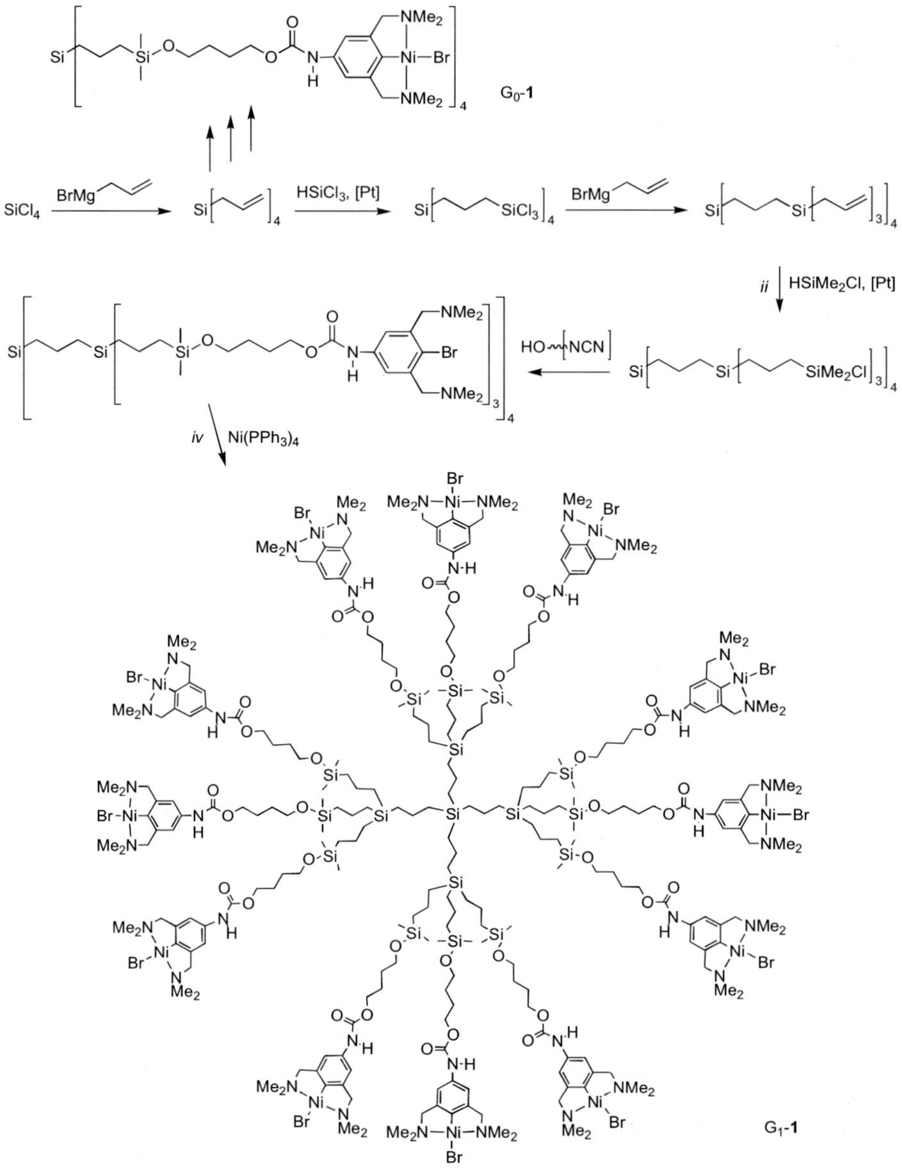

Scheme 2 Schematic synthetic pathway and structure of the carbosilane-based metallo-dendrimers G$_0$-**1** and G$_1$-**1**

study was to vary the density of the catalytic units at the periphery of the dendrimer species through the introduction of different "spacers" in the dendritic framework via different substitution patterns at the branching points (Fig. 4).

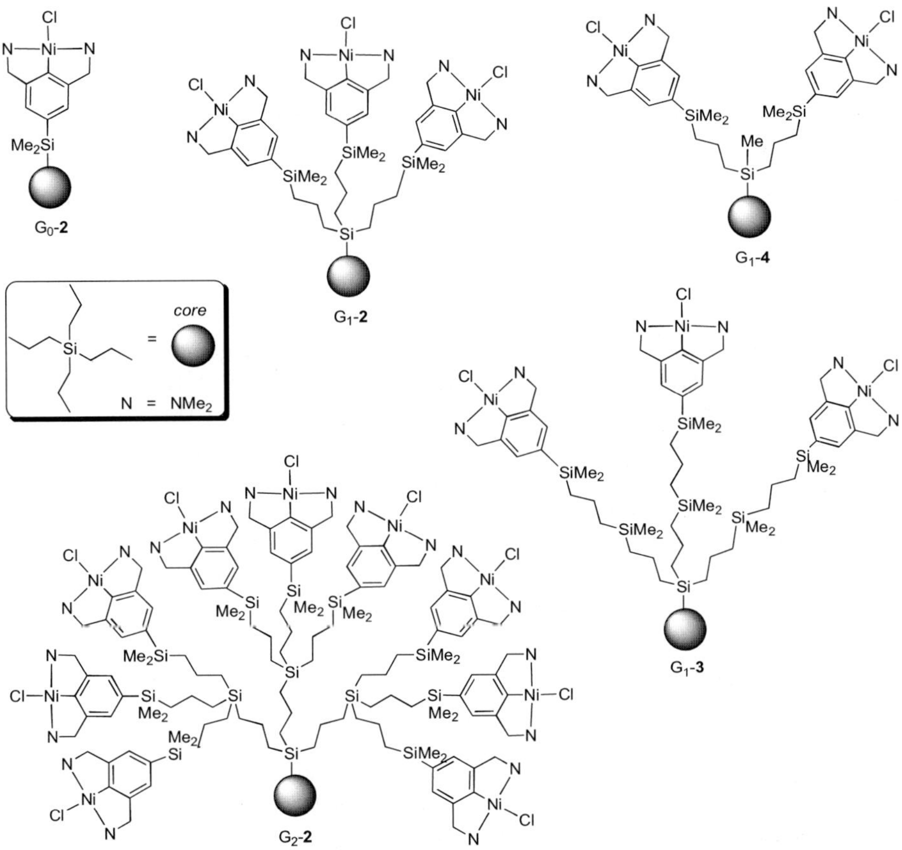

Fig. 4 Schematic structure of parts of Van Koten's carbosilane-based catalysts (note that the structures have four times the number of branches shown here for each dendrimer)

These metallodendrimers were synthesized starting from a chloro-functionalized carbosilane dendrimer, which was used to quench a para-lithiated NCN-pincer ligand. After selective lithiation between the NMe$_2$-arms of the NCN-pincer ligand with t-BuLi, the nickel centers were introduced via a transmetallation reaction with NiCl$_2$(PEt$_3$)$_2$, affording the Ni-containing catalysts (Scheme 3).

Although these synthetic routes are effective, some problems were observed in the last reaction step. Spectroscopic and elemental analysis indicated that the nickellation of the pincer moiety was incomplete, giving an average of 80 to 90% of metallated pincer sites per dendrimer. This observation was rationalized by partial hydrolysis of the reactive lithiated species prior to the introduction of the nickel reagent, causing incomplete metallation of the ultimate dendrimer species [37, 38].

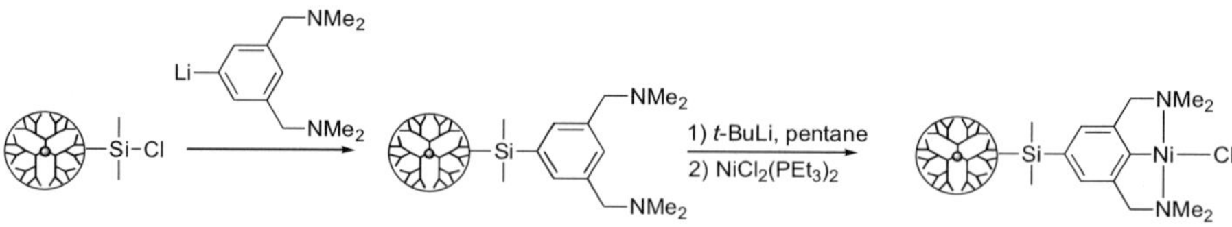

Scheme 3 Schematic synthetic pathway for Van Koten's Ni-based carbosilane metallodendrimers

Applying these catalysts in the Karasch addition reaction, a remarkable degree of variation in activity was observed. The zeroth-generation metallodendrimer G_0-**2** exhibited similar activity in comparison to the monomeric analogs, the first-generation metallodendrimer G_1-**2** (with three metallated pincers per silicon branch, giving in total twelve Ni-containing pincer functions at the periphery of the dendrimer) exhibited an activity of approximately half of that of the G_0-**2** with a significant catalyst deactivation after one hour (characterized by the formation of a purple precipitate), whereas the second-generation metallodendrimer G_2-**2** was essentially inactive. On the other hand, G_1-**3** and G_1-**4**, in which the branches are enlarged and the concentrations of the nickel moieties are diluted, respectively, did not show any sign of catalyst degradation and full substrate conversion was achieved within 22 hours.

Taking into account the mechanism of the Kharasch addition catalyzed by NCN-pincer Ni-complexes, these results were rationalized in terms of deactivation of the Ni(II)-catalyst due to, for example, homocoupling between the $^{\bullet}$CCl$_3$ radicals in the cases where high local concentrations of radicals were generated, i.e. in catalysts with a high concentration of Ni centers on the surface of the dendrimer (G_1-**2** and G_2-**2**). In other words, the irreversible formation of the Cl$_3$C–CCl$_3$ product traps the Ni(III) species and so the active Ni(II) center cannot be regenerated (Scheme 4). By diluting the Ni concentration on the periphery or by extending the internickel distances (as is the case for G_0-**2**, G_1-**3** and G_1-**4**), the local radical concentration is drastically reduced as is the rate of catalyst deactivation (the rate of radical (homo)coupling reactions).

This result, caused by the "proximity effect" between peripheral catalytic sites, can translate into higher or lower catalytic activity of the metallodendrimer in homogeneous catalysis, and is commonly termed the *dendritic effect*. In the above case, a negative dendritic effect is observed. An interesting example of a positive dendritic effect on catalyst activity was reported by Jacobsen et al. in the hydrolytic kinetic resolution of terminal epoxides by peripherally Co(salen)-substituted PAMAM dendrimers [39].

The pincer-based carbosilane dendrimers G_0-**2** and G_1-**2** were tested for their degree of retention in a membrane reactor equipped with a SelRO-MPF-50 nanofiltration membrane [35, 36]. Their retentions were measured

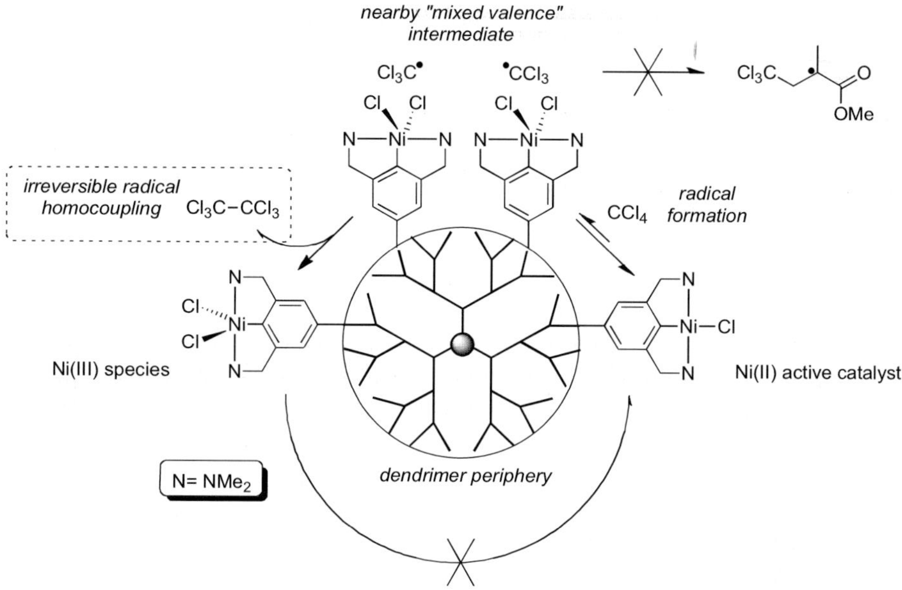

Scheme 4 Proposed deactivation pathway for G_1-2 and G_2-2 catalysts in the Kharasch addition

to be 97.4% for G_0-2 and 99.75% for G_1-2, indicating that the larger dendrimer (G_1-2) is sufficiently retained. Therefore, it is possible to use a continuous flow membrane reactor (CFMR). Applying the G_1-2 dendrimer in a CFMR, a significant loss of catalytic activity was observed over 33 hours. After continuous membrane reactions (after the reactor volume was replaced 64 times), tests of the retained and filtered fractions indicated that the membrane retained the catalyst at 98.6% (which closely resembled the measured retention found earlier in batch-wise processes for this catalyst). Although this degree of retention leads to some loss of catalytic material in the continuous membrane reactor, it could not solely explain the observed deactivation of G_1-2 in the reactor. It was, therefore, proposed that the reactive radical intermediates might be interacting with the functional groups of the membrane material and thus influencing the overall reactivity. While the catalyst was effectively retained, new membranes compatible with the reaction conditions used are needed to successfully construct a useful system for continuous operation.

The synthesis of dendritic carbosilanes functionalized with various diphenylphosphino carboxylic acid ester endgroups has also been reported by the Van Koten group in collaboration with Vogt et al. [40, 41]. The coupling of carbosilane supports containing benzylic alcohol moieties with phosphinoxy carboxylic acid chlorides resulted in the formation of G_0 and G_1 phosphine oxides, which subsequently were converted into the phosphino compounds by reduction with trichlorosilane in benzene. These dendritic hemilabile ligands were applied in the palladium-catalyzed hydrovinylation of styrene. The dendritic catalysts were prepared in situ by treatment of the ligands with $[(\eta^3\text{-}C_4H_7)Pd(cod)]BF_4$ (Scheme 5).

It was shown earlier that palladium-catalyzed hydrovinylation of styrene using phosphino ester-type ligands leads to isomerization of the external alkene (kinetic product) to the internal alkene (thermodynamic product) at higher substrate conversion. In this regard, the idea was to suppress this isomerization by running the reaction at lower conversion in a CFMR system in order to minimize the catalyst–substrate contact time.

Tests made in a batch-wise system showed that G_0-5 had a lower activity than the corresponding monomeric phosphino ester palladium complex. Although less active, the metallodendrimer G_0-5 displayed higher stability

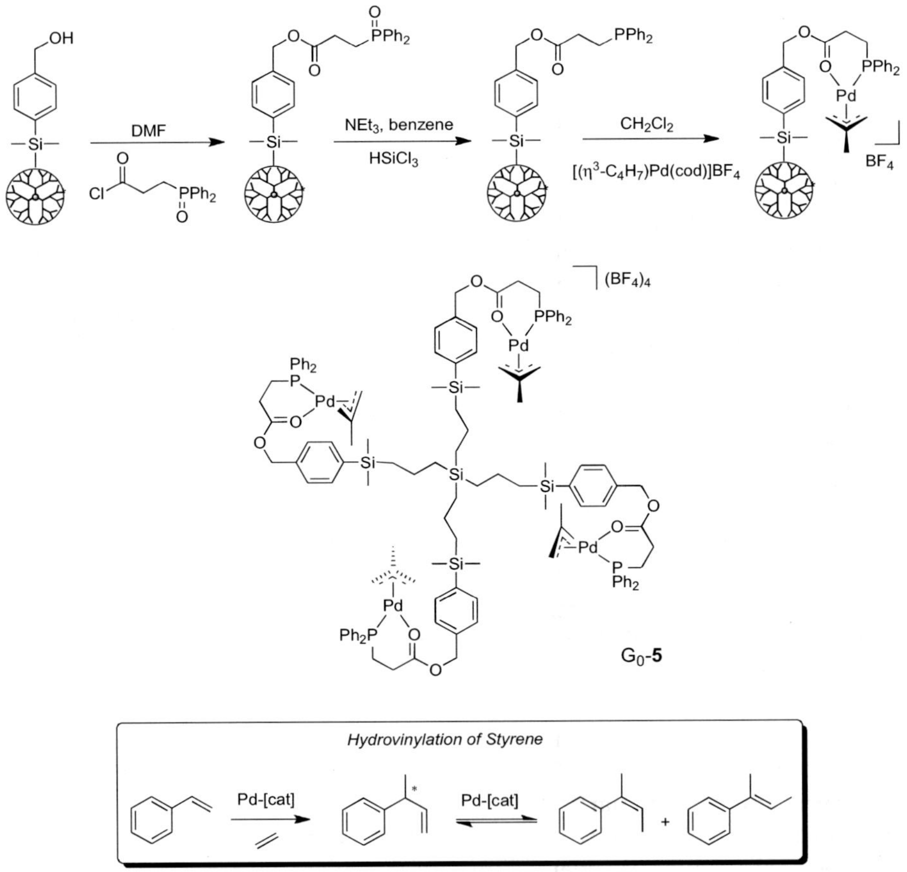

Scheme 5 Schematic synthetic pathways and structure of dendritic P,O ligands applied in the palladium-catalyzed hydrovinylation of styrene

than its parent unsupported ligand catalyst under similar reaction conditions. These features could fit perfectly for its application in a continuously operated high-presssure membrane reactor. In this case, its lower activity together with the short catalyst–substrate contact time may give rise to preferential formation of the kinetic product, while minimizing isomerization.

In a continuous set-up using G_0-5, a considerable decrease in activity was also observed in comparison with the monomeric models. This effect was partly explained by deactivation of the catalyst due to the reaction of the catalytic units with the membrane or to some dendritic effect (due to the proximity of the unit centers leading to double or multiple phosphine complexation with the same palladium center), since the formation of a palladium black precipitate was observed on the membrane. Tests with the G_0 dendrimer model ligand (without palladium) showed a retention degree of 85%. This fact could also partly explain the decrease in activity due to the washout of catalyst during the reaction. However, a first-generation G_1-5 catalyst, with a higher retention, showed almost the same deactivation behavior. Thus, catalyst decomposition is most probably the main reason for the observed deactivation.

Although the retention of the G_0-5 catalyst is modest for practical purposes and the yields obtained are quite low comparing to the monomeric model, the results obtained were promising since no isomerization or formation of other side-products was detected in the product solution.

The first examples of metallated phosphine-functionalized dendrimers that were suitable as homogeneous catalysts were reported by Reetz et al. [42]. DAB-based phosphino dendrimers were synthesized from the commercially available DAB-polypropylene imine dendrimers (DAB = 1,4-diaminobutane) via a double phosphination of the amines with diphenylphosphine and formaldehyde. These dendrimers were loaded with Pd(II) centers by the reaction of the phosphine-functionalized dendrimer with $[Pd(Me)_2(tmeda)]$ (tmeda = tetramethylethylenediamine), in which the diphenylphosphino end groups act as a bidentate ligand (Scheme 6).

Kragl and Reetz demonstrated for the first time the viability of the application of transition metal-catalyzed homogeneous reactions with a dendritic catalyst in a membrane reactor [43]. Indeed, these palladium-based metallodendrimers were applied in the allylic substitution reaction of 3-phenyl-2-propenylcarbonic acid methyl ester with morpholine in a continuous membrane reactor. The application of catalyst G_3-6 for a period of 100 *residence times* (number of replacements of the reaction volume) resulted in a decrease in conversion from 100% to 80%. A small amount of palladium leaching was observed, which could partially explain the decrease in conversion. However, the retention of the supported catalyst (independently determined) was found to be 99.9%, which is considerably higher. Therefore, it was suggested that the drop in conversion might be due to metal–ligand dissociation (i.e., catalyst deactivation or decomposition) rather than leaching of the catalyst.

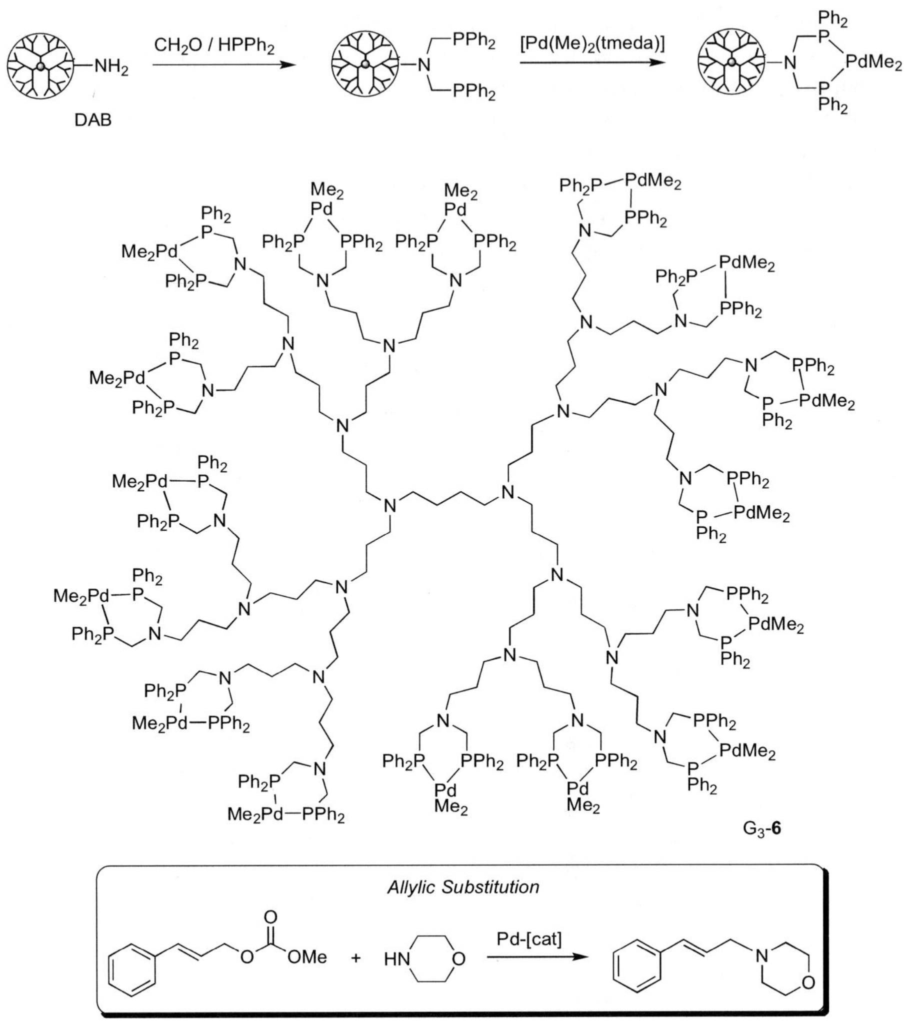

Scheme 6 Reetz's Pd(II) phosphine-functionalized dendrimer applied in allylic substitution

The group of Van Leeuwen has reported the synthesis of a series of functionalized diphenylphosphines using carbosilane dendrimers as supports. These were applied as ligands for palladium-catalyzed allylic substitution and amination, as well as for rhodium-catalyzed hydroformylation reactions [20, 21, 44, 45]. Carbosilane dendrimers containing two and three carbon atoms between the silicon branching points were used as models in order to investigate the effect of compactness and flexibility of the dendritic ligands on the catalytic performance of their metal complexes. Peripherally phosphine-functionalized carbosilane dendrimers (with both monodentate and bidentate phosphine endgroups) were synthesized by reacting different generations of chlorosilane dendrimers with the tmeda complex of [(di-phenylphosphinyl)methyl]lithium (Scheme 7).

Palladium complexes with monocoordinated phosphine ligands were synthesized by mixing the corresponding phosphine dendrimer with [Pd(cod)MeCl], which afforded exclusively *trans* dendrimer(MePdCl)$_n$ complexes (P/Pd = 2). Complexes in which the ligand is coordinated in a bidentate fashion were synthesized by reaction of the respective bisphosphine den-

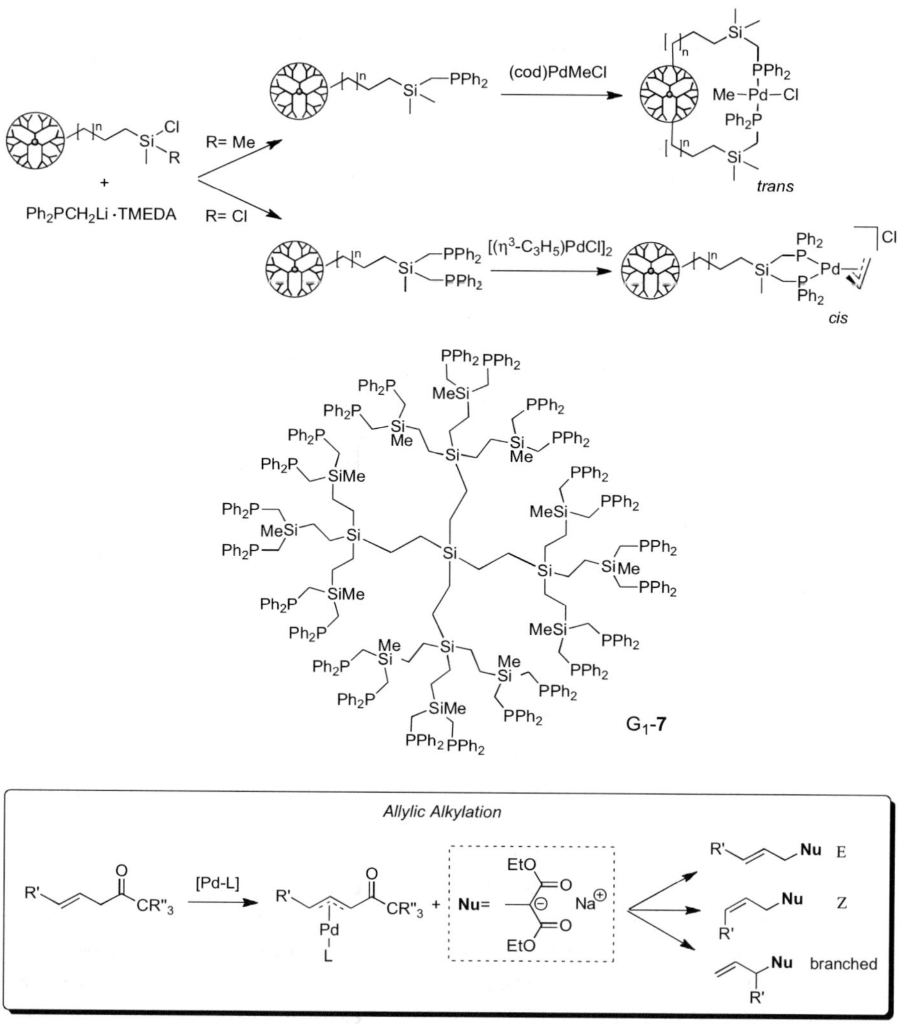

Scheme 7 Schematic pathway of the synthesis of diphenylphosphine-functionalized carbosilane metallodendrimers applied in allylic alkylation reactions

drimers with $[(\eta^3\text{-}C_3H_5)PdCl]_2$. These (allyl)palladium catalysts were applied in the reaction between sodium diethyl 2-methylmalonate and allyl trifluoroacetate ($R' = $ Me, Ph; $R'' = $ F) in batch reactions and CFMR set-ups [44, 45].

In batch processes, the monodentate catalysts showed lower activity compared to their bidentate analogs. The activity per palladium center was constant upon increasing the dendrimer generation of the dendritic Pd(allyl) complexes, indicating that all active sites act as independent catalysts. In addition, the selectivity between the E- and Z-products was similar to that induced by analogous mononuclear palladium complexes. Although a considerable amount of the branched product was observed, the authors did not put forward an explanation for its formation.

Under continuous reaction conditions, however, using the largest ligand G_1-7 ($R > 98\%$), an unexpected rapid decrease in activity was found that could not be solely ascribed to washout of the homogeneous catalyst, but was proposed to be a consequence of catalyst decomposition during the process.

The same dendritic ligands, but used in combination with rhodium, were utilized in hydroformylation reactions [46]. Preliminary experiments with this catalytic system in a nanofiltration membrane reactor, however, showed that this membrane set-up was not compatible with the standard hydroformylation conditions because of its temperature and solvent restrictions.

Olefin metathesis has become a very important reaction in polymer chemistry and natural product synthesis [47–49]. Garber et al. have used the physical properties of dendrimers in order to improve the separation between the dendritic metathesis catalyst and products on silica gel column chromatography [50]. The Van Koten group has reported on the synthesis of different generations of carbosilane dendrimers functionalized with ruthenium metathesis catalysts [51].

The synthetic method used to arrive at these systems starts with the quenching of *para*-bromo-lithiobenzene with Me_2SiCl-terminated carbosilane dendrimers, affording the 4-bromobenzene-functionalized dendrimers (Scheme 8). These compounds were again lithiated with *n*-BuLi via lithium-halide exchange and the resulting products were immediately reacted with 2-acetylpyridine, affording the 2-hydroxyalkylpyridyl-functionalized dendritic compounds. Finally, deprotonation of these precursors with *n*-BuLi and subsequent transmetallation with $PhC(H) = RuCl_2(PR_3)_2$ afforded the ruthenium metathesis dendrimers.

The ruthenium-based metallodendrimer G_0-8 was applied as catalyst in a ring-closure metathesis reaction. The activity per metal center of the dendritic catalysts was found to be comparable to that of the corresponding mononuclear catalyst. Unfortunately, the metathesis reaction conditions were not compatible with the nanofiltration membrane set-up used, since a black precipitate was formed in the vessel containing the catalyst. It was found that the conversion to diethyl-3-cyclopentene dicarboxylate product stopped

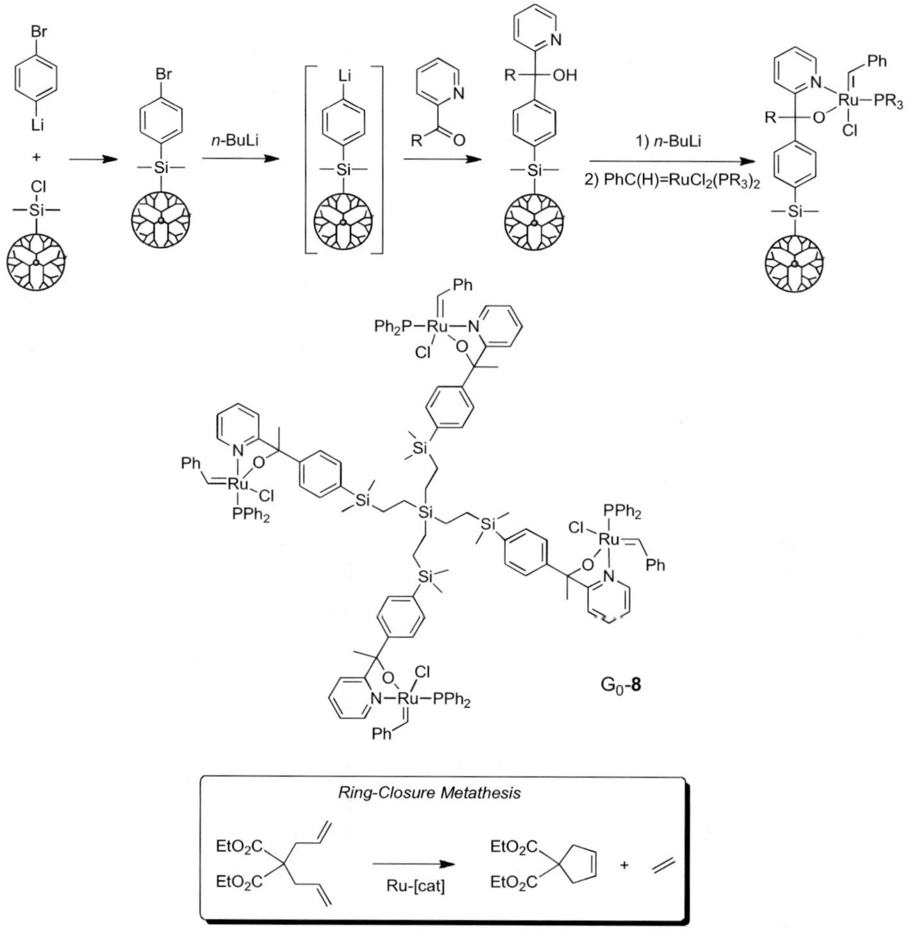

Scheme 8 Synthesis of Van Koten's dendritic ruthenium metathesis catalyst applied in ring-closure metathesis

after 20% conversion. This phenomenon was proposed to be due to interaction of the catalyst with the membrane surface and not to leaching of the catalyst.

Togni et al. have attached chiral ferrocenyl diphosphines (Josiphos) onto the periphery of a dendritic skeleton via a silane–carbon linkage with the carbon of the cyclopentadienyl ring of the ferrocene (Scheme 9) [52–55]. This Josiphos-containing dendron was connected to different cores, and metallodendrimers containing up to 24 metallocene fragments were synthesized. The synthetic procedure is essentially the same for all of the dendrimers and consists of a coupling between the phenol-based dendron with oligo-carboxylic acid cores.

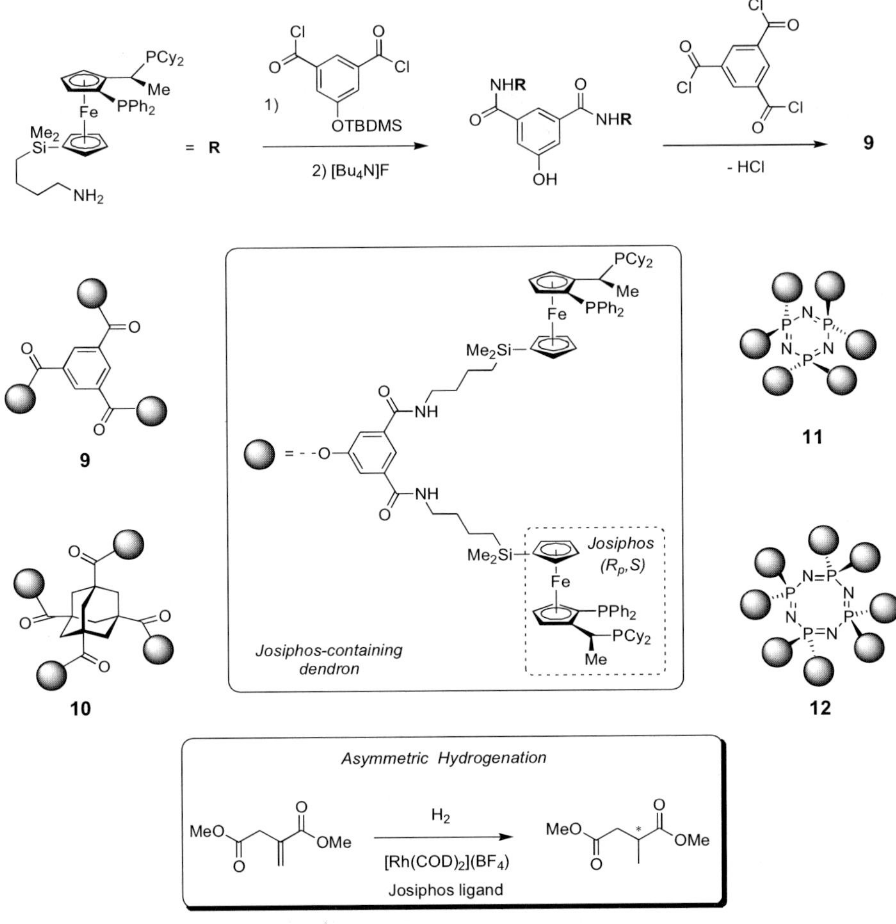

Scheme 9 Structure of Togni's Josiphos-containing metallodendrimers applied in rhodium-catalyzed asymmetric hydrogenations

These Josiphos ligand systems have been proven to be effective catalysts for Rh-catalyzed asymmetric hydrogenation of alkenes. Indeed, by reacting these dendrimers with one equivalent of $[Rh(COD)_2]BF_4$ (per Josiphos) in CH_2Cl_2, the active catalytic species is generated in situ after 15 min and hydrogenation of dimethyl itaconate is complete after 20 min. In all cases, Togni's dendritic systems exhibited enantiomeric excesses (ee) of more than 98%, which is similar to the ee obtained by a monomeric model applied in an identical reaction. These results showed that the chiral peripheral organometallic dendrimers can be efficiently applied in enantioselective catalysis and, considering the different sizes, numbers of Josiphos groups and dendrimer generations of the catalysts used, that each catalytic center is essentially independent of its neighbors.

For the larger dendrimers incorporating an arene **9** and adamantyl core **10**, preliminary membrane filtration experiments showed quantitative retention of the dendrimer–rhodium complexes using a Millipore Centricon-3 membrane and methanol as the solvent [53]. Results on hydrogenation reactions using these dendrimer catalysts in a CFMR have not yet been reported.

Peripheral ferrocenyl-functionalized dendrimers are in fact one of the most common type of metallodendrimers synthesized today. Due to the chemical and thermal stabilities of this kind of metallocene, formation of the intended dendrimer-supported metallocene complex can be performed in a single reaction step: the metallocene fragment is introduced onto the surface of a (commercially available) dendrimer as the final construction step [56, 57].

Astruc et al. [58] have employed noncovalent hydrogen bonding as a synthetic pathway in order to construct a dendrimer containing ferrocenyl groups. Starting from commercially available DAB-poly(propylene imine)dendrimers of generations 1–4 and dendrons containing ferrocenyl moieties and one phenolic group, the driving force for the formation of the dendrimer assemblies are the $(H_2N) - (HO)$ hydrogen bonding interactions (Fig. 5). The presence of hydrogen bonding interactions was established by the chemical shift and shape of both NH_2 and ArOH proton signals in NMR experiments. These ferrocenyl-substituted H-bonded dendrimers were used for the recognition of $H_2PO_4^-$ anions, which was probed by cyclic voltammetry studies [59]. Although not applied in catalytic reactions, the principle demonstrated by Astruc has been widely used to build up new (metallo)dendritic structures through self-assembly.

The Van Koten–Klein Gebbink group has used electrostatic interactions in a similar fashion to the H-bonded dendrimers of Astruc (Fig. 5), as the key to incorporating organometallic NCN-pincer fragments into a dendritic framework [60, 61]. Utilizing NCN-pincer palladium complexes in which a sulfato group is tethered via an alkyl chain to the *para*-position of the pincer aryl ring, up to eight pincer moieties have been introduced into a dendrimer containing an octacationic ammonium core and a variety of polar or apolar dendritic arms (Scheme 10). The byproduct of the synthesis, $[NBu_4]Br$ (the NBu_4 cation from the organometallic fragment and the bromide anion from the dendrimer), could be removed by washing the product with water. In order to completely purify the metallodendritic assembly (to remove access of the sulfato-NCN-pincer Pd-complex), a passive membrane dialysis experiment was performed to get rid of the unsupported NCN-pincer palladium starting material.

The neutral and cationic [Pd(II)NCN] pincer-containing dendritic assemblies **14** and **15** have been used as Lewis acid catalysts in the aldol condensation of benzaldehyde and methyl isocyanoacetate. These showed conversions, turnover frequencies and *cis/trans* ratios of the oxazoline products that were essentially identical to those of the monomeric model complex applied in the same reaction.

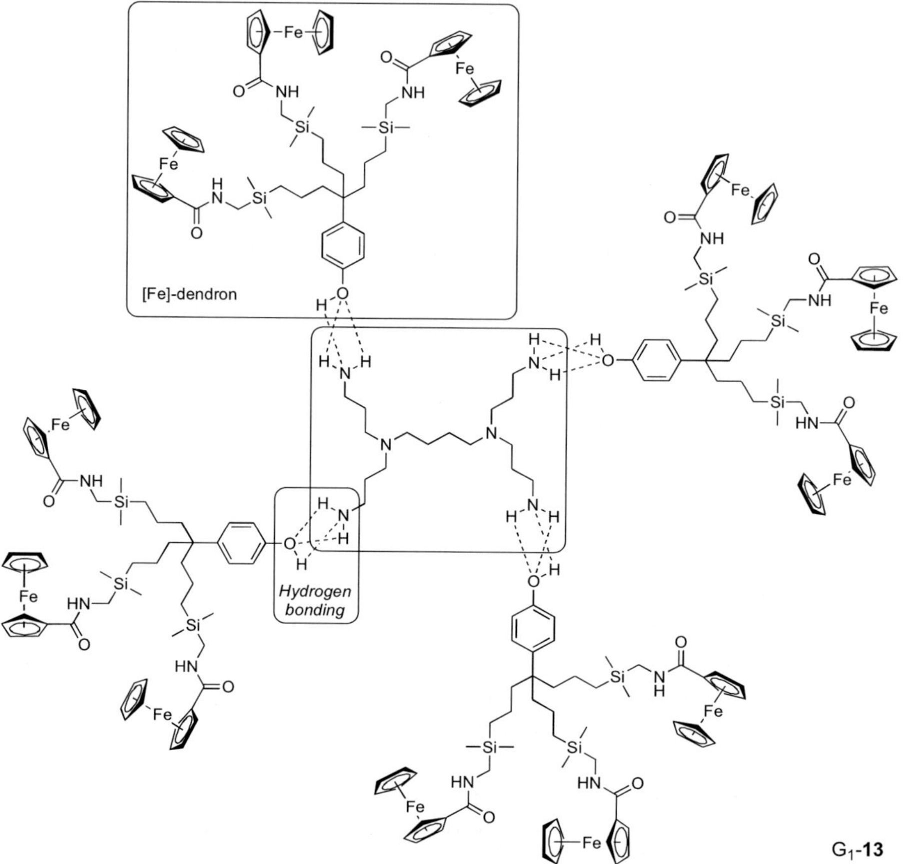

Fig. 5 Astruc's first-generation ferrocenyl-substituted dendrimer

Although CFMR experiments have not yet been performed with these metallodendritic assemblies, their purification using passive dialysis showed the potential application of these catalysts in a recycling process by means of membrane filtration techniques.

The groups of Reek and Meijer have also applied the noncovalent approach for catalyst anchoring to a dendrimer support [62]. Phosphine functionalized ligands were attached to the periphery of poly(propylene imine) dendrimers via combined ionic interactions and H-bonding using a specific binding motif that is complementary to that of the support (Fig. 6).

The resulting peripheral phosphine-functionalized dendrimers were used as a ligand in the palladium-catalyzed allylic amination of crotyl acetate in a CFMR. The active catalysts were prepared by mixing all three components: the dendrimer, the phosphine ligand and a suitable palladium precursor [(crotyl)PdCl]₂. The catalytic activity and selectivity of **16** in a batch process

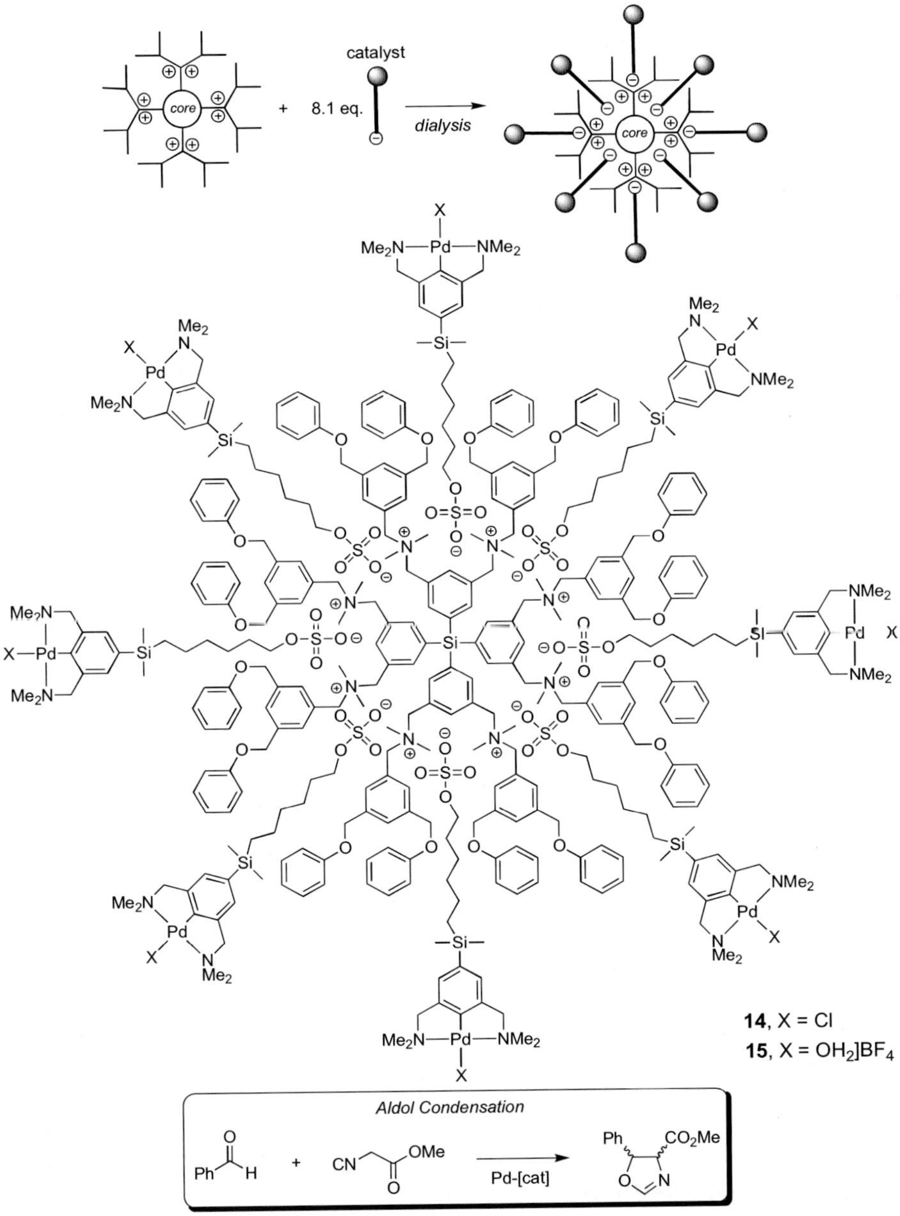

14, X = Cl
15, X = $OH_2]BF_4$

Scheme 10 Structure of the Van Koten–Klein Gebbink noncovalent dendritic catalyst assemblies applied in aldol condensation reactions

was found to be similar to that of a mononuclear model compound. Membrane retentions of 99.4–99.9% were found to be dependent on the P/Pd atomic ratio used. A small decrease in conversion was observed when the

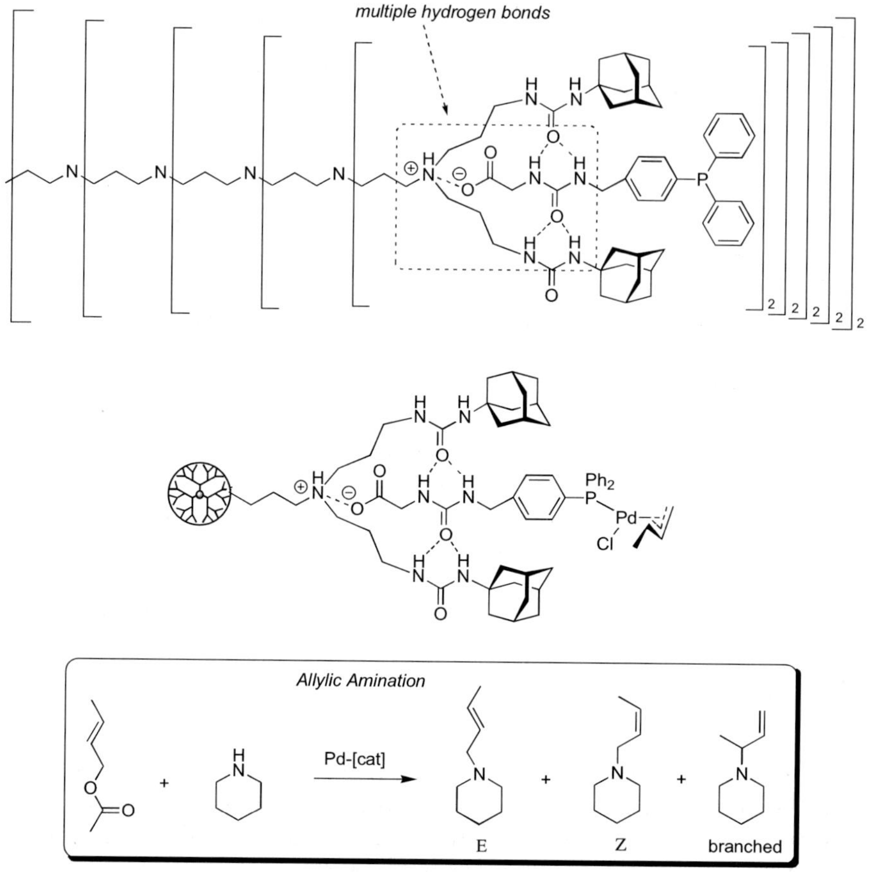

Fig. 6 Schematic representation of the Meijer–Reek noncovalently immobilized phosphine ligands applied in palladium-catalyzed allylic aminations

dendritic catalyst was applied in a CFMR. This decrease could not completely be explained by washout of the catalysts, but was proposed to be caused by deactivation of the catalyst.

4.2
Core-Functionalized Catalysts

The first example of the integration of a molecular catalyst into the core position of a dendrimer was reported by Bruner et al., who studied the influence of a chiral dendritic periphery on the performance of cyclopropanation catalysts [63]. Ever since, a series of reports on the application of chiral core-functionalized metallodendrimers in asymmetric catalysis have appeared [64–66]. Among these catalysts there are few examples of the use of nanomembrane filtration techniques for catalyst recycling.

Ferrocenyl diphosphine core-functionalized carbosilane dendrimers have been prepared as ligands for homogeneous catalytic reactions applied in a CFMR by Van Leeuwen et al. [20, 21, 67, 68]. The syntheses of these *dppf*-like ligands (G_0–G_2)-**17** were performed using carbosilane dendritic wedges with an aryl bromide as focal point. These wedges were coupled to the core via quenching of the lithiated species with ferrocenyl phosphonites (Scheme 11).

Dendrimers containing allyl moieties as peripheral groups (R = allyl, Scheme 11) were used as ligands in palladium-catalyzed allylic alkylation (see the reaction in Scheme 7). Their reaction with [$PdCl_2(MeCN)_2$] was monitored by ^{31}P NMR spectroscopy, which confirmed the complete complexation of the ligand in a similar way as observed for the analogous monomeric dppf complexes; in other words the phosphines coordinate to the metal in a bidentate *cis* fashion. In order to perform the catalytic reactions, the active palladium catalysts were prepared in situ by reacting the ligands with crotyl-palladium chloride [(η^3-C_4H_7)PdCl]$_2$ (P/Pd = 2) in THF followed by addition of the reactants.

A similar conversion was found for dendrimers **17** and the analogous monomeric dppf, indicating that the allyl end groups did not interfere in the catalytic reaction. The activity and regioselectivity of the alkylation reaction changed according to the dendron generation. When higher generations or larger dendritic ligands were applied (from G_0-**17** to G_2-**17**, respectively) a decrease in overall catalytic activity and an increase in the formation of the branched allylation product were observed. These observations were attributed to the lower accessibility of the catalytic site located at the core of the dendrimer. In other words, the increased steric bulk of the dendrons embracing the catalytic site hinder the attack of the nucleophile (sodium diethyl 2-methylmalonate) on the allyl–Pd–dendrimer species.

The applicability of this type of dendritic catalyst in a CFMR was tested for G_2-**17**. It was observed that the catalytic activity remained almost constant for up to eight hours of reaction time, which is in contrast with the peripheral-functionalized catalysts (G_1-**7**, Scheme 7) applied under similar conditions. Although resulting in an overall lower activity per catalytic center, the location of the catalytic site within the dendritic sphere seems to protect the active species against deactivation via interaction with the membrane or with other metallodendritic species.

Diphosphine core-functionalized dendrimer analogs containing a methyl moiety as the peripheral group (R = Me, Scheme 11) have been applied as ligands in rhodium-catalyzed hydroformylation and hydrogenation of alkenes. For the hydrogenation of dimethyl itaconate (Scheme 9), the rhodium catalyst was formed in situ via the reaction of [$Rh(nbd)_2$]$_2$(ClO_4) (nbd = 2,5-norbornadiene) with the diphosphine dendritic ligand (P/Rh = 2) [68]. All dendritic dppf-type ligands gave active hydrogenation rhodium catalysts with

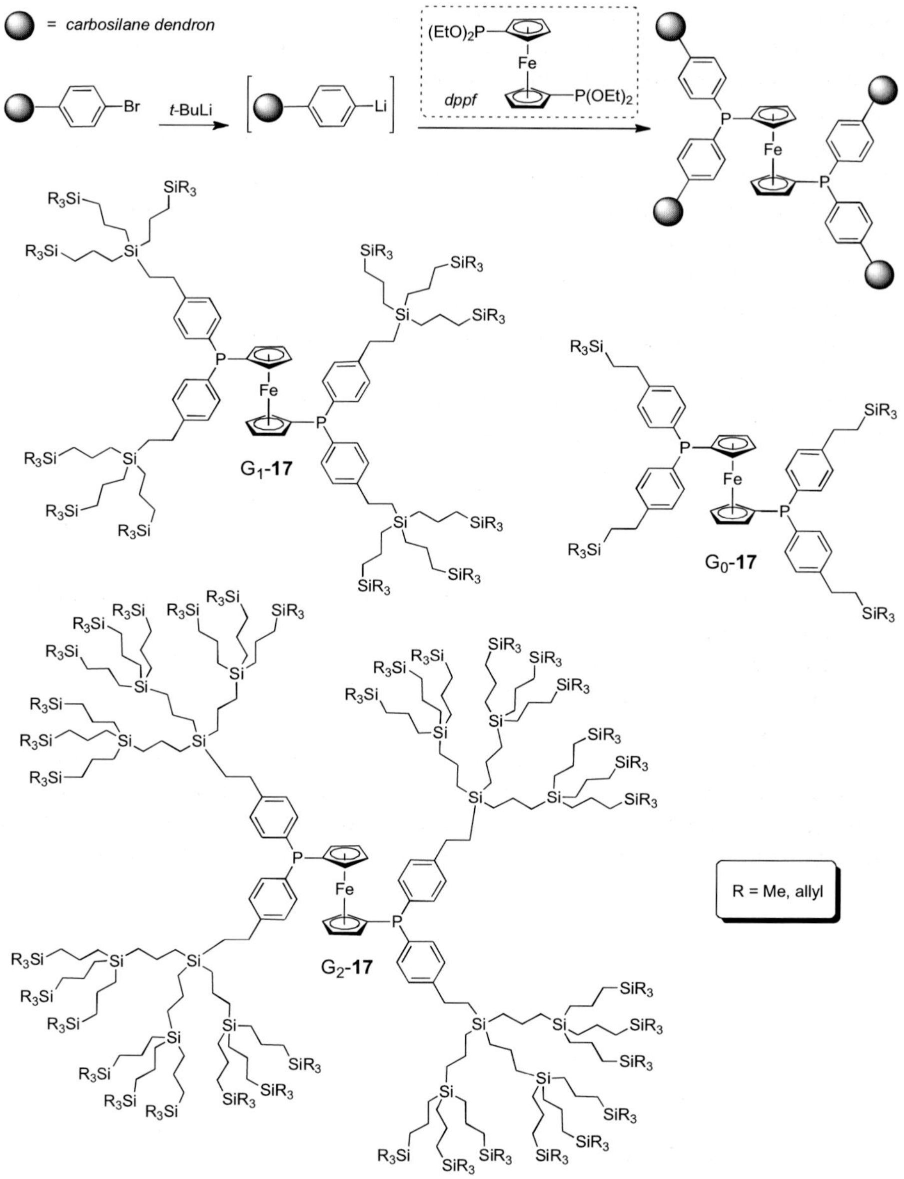

Scheme 11 Van Leeuwen's core-functionalized ferrocenyl-phosphine carbosilane dendrimers used as ligands in palladium-catalyzed allylic alkylation reactions

activities similar to the analogous monomeric dppf ligand under the same conditions. In a CFMR, however, the reaction applying the rhodium catalyst of dendrimer G_1-17 (R = Me) gave a higher maximum conversion and was more stable over time (77–85% of conversion for 35 reactor volumes) than dppf itself (15–70% of conversion for 35 reactor volumes). The lower conversion observed with dppf used as the ligand was attributed to its lower rate of hydrogenation compared to G_1-**17** and to leaching of the active Rh–dppf complex.

ICP–AES analysis (induced coupled plasma atomic emission spectroscopy) was performed on the reactor permeate, which showed that the amount of metal leaching is similar to that of the phosphine ligand. This indicated that the ligand–metal interaction was sufficiently strong under the applied reaction conditions. In addition, measurements on the parent ligands in the membrane reactor showed a retention of 87.5% for dppf and 99.4% for G_1-**17**, indicating a significantly enhanced retention behavior of the "dendronized" dppf ligands.

The Van Koten group has developed an interesting approach to the assessment of the permeability of nanofiltration membranes for the application of metallodendrimer catalysts in membrane reactors. They have selectively grafted dendrons to organometallic pincers with sensory properties and have used these as dyes in a colorimetric monitoring procedure.

Square-planar platinum or nickel complexes derived from the NCN-pincer ligand framework react selectively and reversibly with SO_2 to form a pentacoordinate adduct $[MX(NCN)(SO_2)]$ via m-S η^1 bond formation (Scheme 12). Upon the binding of SO_2, the organometallic species undergoes a reversible color change, from colorless to bright orange in the case of NCN-pincer platinum halide complexes. NCN-pincer platinum complexes can therefore be used as a diagnostic agent or sensor for SO_2 gas and vice versa [14–18].

This sensory property was used to probe the suitability of metallodendrimers for nanofiltration membrane techniques in homogeneous systems. During continuous-flow membrane filtration, any leaching of a metalloden-

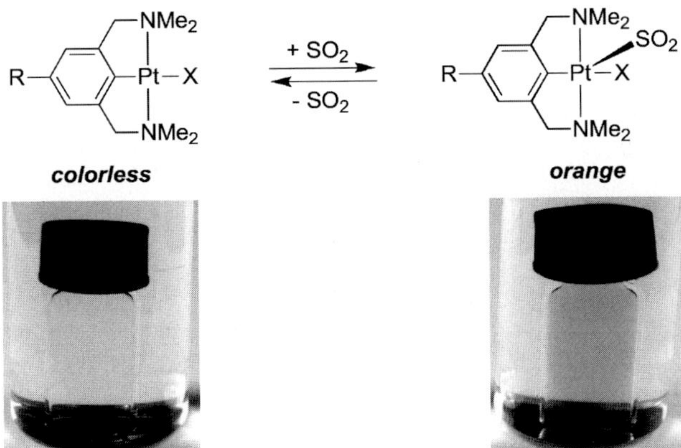

Scheme 12 Color change in a solution of the NCN-pincer platinum complex upon reversible binding with SO_2 gas (R = H, Fréchet dendron; X = halogen); see also Fig. 7

drimer can be directly monitored by treatment of the permeate with SO_2 (monitored on-line by UV–visible spectroscopy).

For this purpose, a series of Fréchet-type polyether dendrons of zeroth- to third-generation containing an NCN-pincer metal complex (M = Pt, Ni) at the focal point of the wedge were prepared [69]. Their synthesis was either

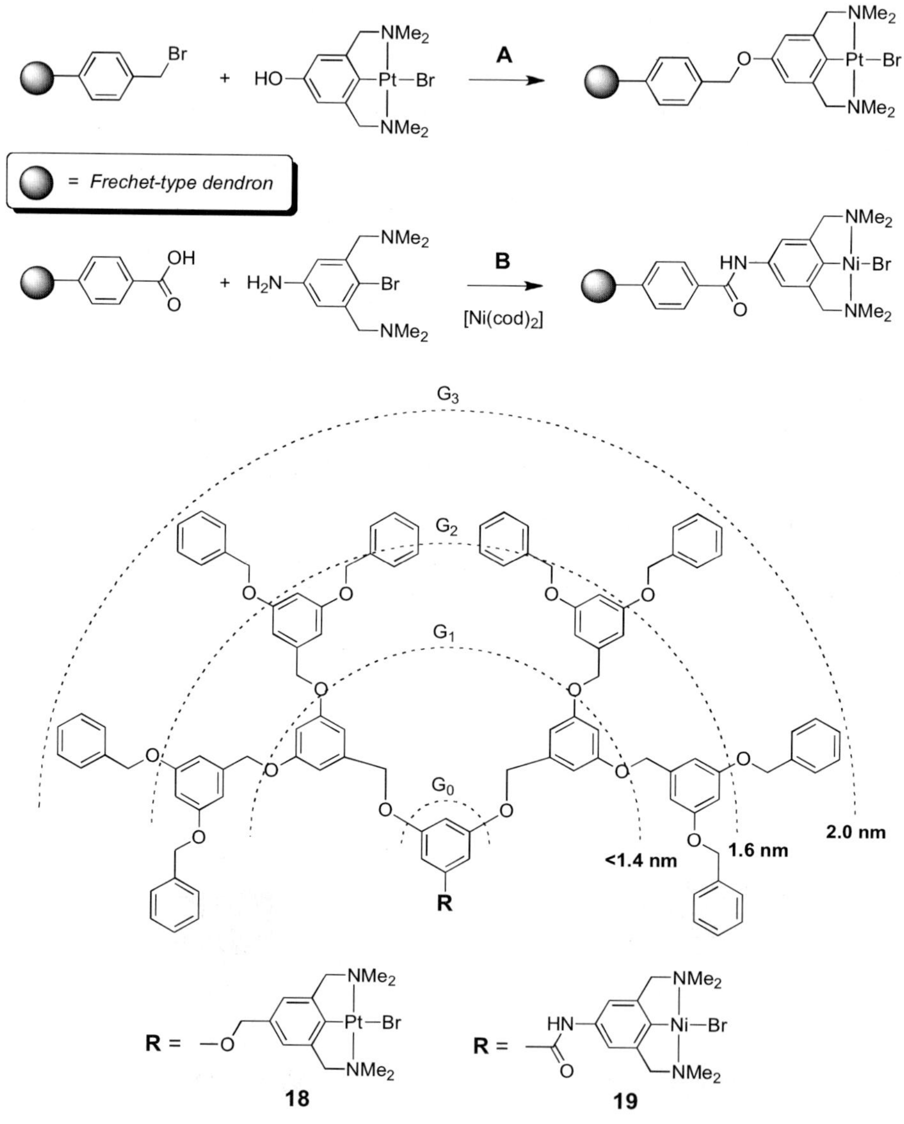

Scheme 13 Schematic representation of Van Koten's NCN-pincer Pt- (path A) and Ni-based (path B) Fréchet-type metallodendrons

achieved by etherification of the phenolic hydroxyl group of an NCN-pincer platinum complex with dendritic benzylic bromides (path A, Scheme 13) or by amidation of a diamino bromo aryl NCN-pincer ligand precursor containing a terminal amine group with a benzylic benzoic acid, followed by oxidative addition reaction with $[Ni(cod)_2]$ (path B, Scheme 13).

Applying the dye approach for metallodendrons (G_1-G_3)-**18**, their retention during nanofiltration showed a linear correlation with their weight and size. An increase in the degree of retention was observed as a function of increasing dendron generation (from G_1- to G_3-**18**, respectively).

Assuming that substitution of the platinum center in G_3-**18** at the core by nickel gives rise to a dendritic catalyst with the same permeability properties (degree of retention), the metallodendron G_3-**19** was loaded in a membrane-covered vial and tested as catalyst in the Kharasch addition of CCl_4 to methyl methacrylate. The compartmentalized catalytic system operates fully homogeneously without catalyst leaching, and the catalyst is separated from the product solution by simply removing the vial (tea-bag approach) and is further purified by washing the catalyst-containing immersion vial with pure solvent (Fig. 7).

In this system, the catalyst G_3-**19** showed a similar reaction rate and turnover number as observed with the parent unsupported NCN-pincer nickel complex under the same conditions. This result is in contrast to the earlier observations for periphery-functionalized Ni-containing carbosilane dendrimers (Fig. 4), which suffer from a negative dendritic effect during catalysis due to the proximity of the peripheral catalytic sites. In G_3-**19**, the catalytic active center is ensconced in the core of the dendrimer, thus preventing catalyst deactivation by the previous described radical homocoupling formation (Scheme 4).

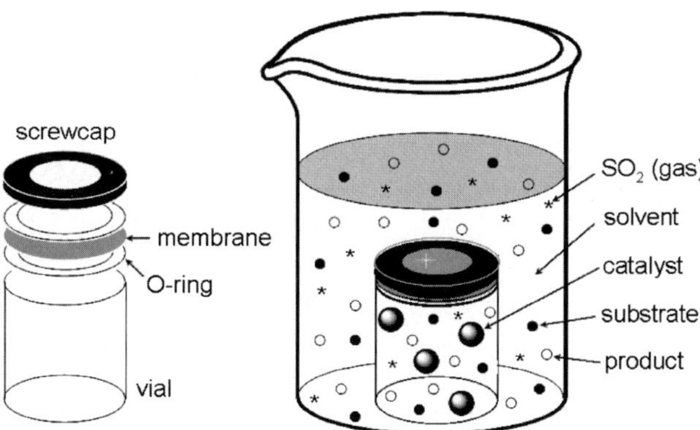

Fig. 7 Schematic representation of compartmentalization of nanosized catalysts in membrane-covered vial reactors

This work points to the use of compartmentalized metallodendrimers as catalysts for continuous flow operations and cascade-type synthetic applications. However, since the driving force in the applied set-up is based on *osmosis (passive diffusion)*, the product flux is limited.

The Van Koten group has also prepared copper-based metallodendritic catalysts via a new convergent synthetic method [70]. Aminoarenethiolate copper(I) complexes (Fig. 8) are catalysts for 1,4-Michael addition reactions among other reactions [71]. These compounds possess interesting features such as robustness, thermal and air stability, good solubility in common organic solvents, and nontransferability of the arenethiolate group (the covalent thiolate–copper bond "ArS – Cu" remains intact throughout the catalytic cycle) [72–77]. Another important characteristic of these complexes is that the endogenous nitrogen coordinates intramolecularly with the metal, while the sulfur atom bridges between two copper centers inducing the formation of aggregates. Both in the solid state and in solution, these species usually exist as trimers in the resting state (Fig. 8) [78–80].

Based on this structural property, the expected formation of aggregates of dendronized complexes of this type would lead to catalysts with appropriate sizes for application in membrane filtration techniques. In this regard, an aminoarenethiolato copper(I) catalyst was attached to the focal point of a small carbosilane dendritic wedge (Scheme 14). The synthesis was performed by lithiation of a *para*-bromo-functionalized aminoarene ligand and subsequent coupling with an appropriated chloro carbosilane wedge. A second lithiation of the supported ligand was accomplished by the addition of a twofold excess of *t*-BuLi to ensure complete lithiation at the position *ortho*

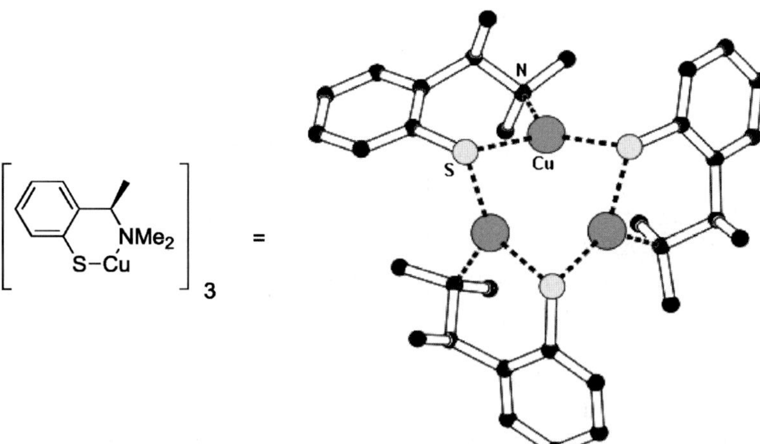

Fig. 8 X-ray structure of the aggregated species formed by an enantiopure (*R*) [α-methyl-(dimethylamino)methylarenethiolate copper(I) complex applied in Michael addition reactions [78–80]

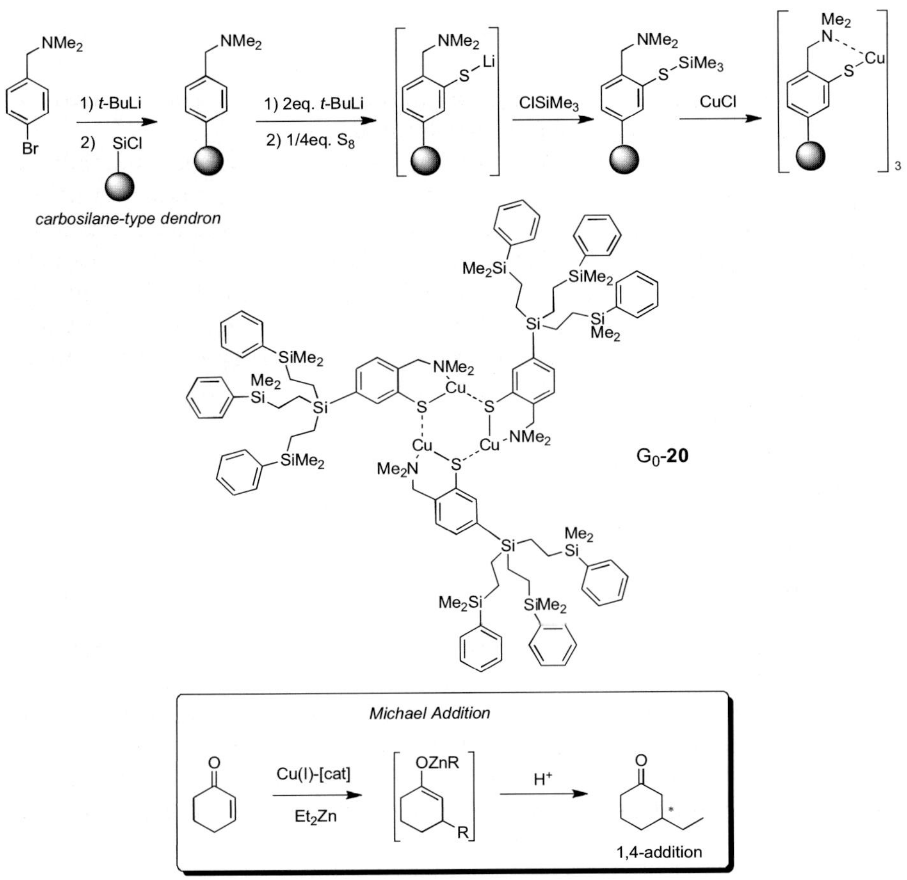

Scheme 14 Aminoarenethiolato copper(I)-based metallodendrimer G_0-**20** and its application as catalyst in a 1,4-Michael addition reaction

to the amino substituent [81, 82]. Then, a stoichiometric amount of sulfur (relative to t-BuLi) was used to accomplish sulfur insertion into the carbon–lithium bond. The resulting lithium thiolate intermediated was quenched with TMSCl, resulting in the TMS-protected dendronized thiolate ligand, which was further reacted with an equimolecular amount of CuCl to afford the desired copper(I)-based metallodendrimer G_0-**20** [70].

Metallodendrimer G_0-**20** was used as a catalyst in the 1,4-addition reaction of diethylzinc to 2-cyclohexenone in a variety of solvents. The results obtained showed that this dendritic catalyst provides activities similar to or even higher than those observed for the unsupported aminoarenethiolato copper(I) complexes, depending on the solvent used. Applying G_0-**20**, this reaction could also be performed using a solvent as apolar as hexane, whereas the unsupported complexes are not soluble in this medium.

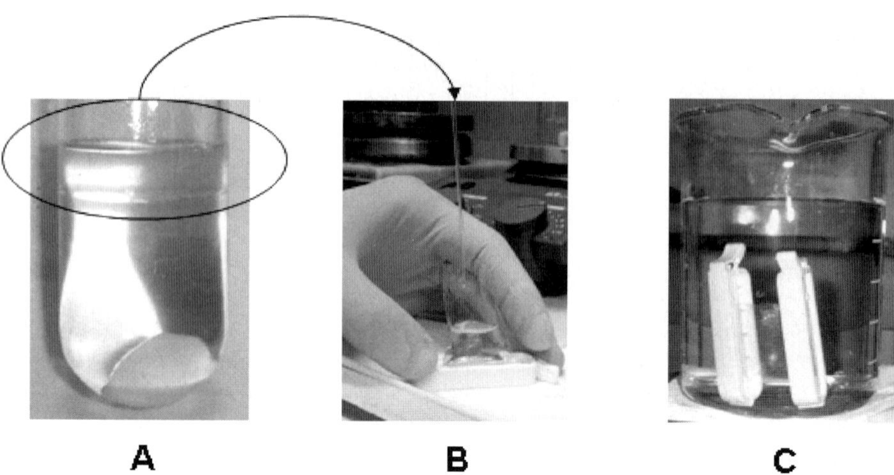

Fig. 9 Passive membrane dialysis performed with catalyst G_0-**20** in a Michael 1,4-addition reaction. **A** Upper phase = Et_2O + catalyst + product, lower phase = aqueous HCl + side-products. **B** Organic phase addition into the "membrane pocket". **C** Diffusion of the product into the beaker (charged with pure Et_2O), while the catalyst remains retained inside the "tea-bag"

At the end of a typical reaction, the reaction mixture was quenched with aqueous HCl in order to form the product (Fig. 9A). The resulting organic phase was submitted to passive membrane dialysis (Fig. 9B,C) to recover the catalyst G_0-**20**. The fraction retained by the membrane (Fig. 9C) was used in another run, where again the 1,4-addition product was formed quantitatively, proving the presence of active catalyst after recycling.

These results showed the high stability of the resting state of G_0-**20**, which most likely originates from the isolating effect on the catalytic site caused by the inert carbosilane dendron. In contrast, the parent unsupported aminoarenethiolato copper(I) complexes decompose rapidly during a similar work-up, as discussed for the supported catalyst. The increased catalytic performance may be attributed to the presence of a bulky carbosilane substituent on the catalyst monomer, which could facilitate dissociation of the resting state aggregates (Fig. 8) to provide monomeric catalyst–substrate as well as catalyst–substrate–reagent intermediates from which product formation occurs.

5
Summary and Outlook

Several examples of transition metal-based catalysts supported via different approaches and located at different positions within various kinds of dendrimers have been developed and applied in homogeneous catalysis. Some of these examples are gathered in this chapter, which provides the overall picture that the macromolecular size and globular shape of a dendrimer support provide homogeneous catalysts that can be recycled by means of membrane filtration techniques.

Nonetheless, the transposition of homogeneous catalytic reactions from unsupported to dendrimer-supported catalysts is still not straightforward. Various "dendritic effects", positive and negative ones, on the activity, selectivity, stability and solubility of metallodendrimer catalysts have been observed in this respect. In our own research we have found that a high concentration of metal centers at periphery-functionalized metallodendrimers may translate into a decrease in the catalytic performance due to undesirable side-reactions between the catalytic sites at the dendrimer "surface" (Fig. 4 and Scheme 4). In contrast, when the exact same catalyst is located at the focal point of a dendron, this matter is avoided by isolating the active site, thereby providing a more stable albeit less active catalyst (Scheme 13).

General problems concerning the application of a metallodendrimer as a recyclable catalyst are related to matters such as dendrimer decomposition, catalyst deactivation or leaching of macromolecular or molecular catalyst species during membrane filtrations. We have investigated one aspect in more detail: the importance of rigidity in the central core of a metallodendrimer on its catalytic performance and degree of retention by nanomembrane filtration [83–88]. Through a study on multimetallic cartwheel-type pincer compounds, such as the one shown in Fig. 10, we have shown that by increasing the size of a material possessing a shape-persistent backbone, a linear increase in the retention of these molecules by nanofiltration membranes is achieved. Moreover, the catalyst with the higher degree of retention and with a high catalytic site concentration at its periphery gave rise to enhanced catalytic activities in comparison to its monomeric counterparts. This example represents an interesting case of a (double) positive dendritic effect. Application of a palladium version of this catalyst in a CFMR produced product for 27 h without showing any change in conversion, selectivity or productivity. An attractive aspect of this kind of soluble, shape-persistent multisite catalyst is the relatively facile synthesis involving short synthetic routes.

In summary, the performance of a recyclable metallodendrimer in homogeneous catalysis depends on different factors, but mainly on its intrinsic chemical and structural properties. At the catalytic level, tailor-made dendritic-supported catalysts can be achieved by choosing an appropriate type of dendrimer backbone, shape and generation and by carefully selecting the locations of and the spacing between the catalytic sites within dendrimer. In this respect, the use of alternative dendritic macromolecular structures as catalyst supports is of interest. Two recent examples from our own work may serve to exemplify this point. In the first example, organometallic moieties were grafted with greater than 93% efficiency onto dendronized polystyrene

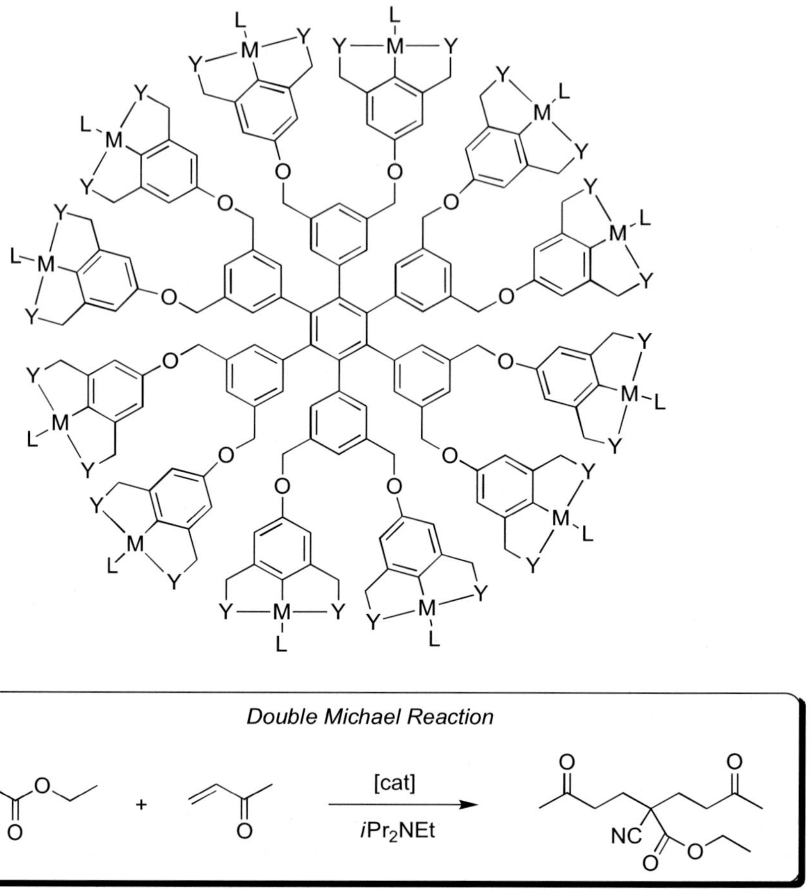

Fig. 10 Shape-persistent nanosized Lewis acid catalyst for double Michael addition reaction

in collaboration with Schlüter et al. (Fig. 11) [89]. In this manner, dendritic macromolecules were obtained that contain 850 (G$_1$), 1700 (G$_2$), and 3400 (G$_3$) identical catalytic Pd sites per molecule. The physical properties of these structures allow for their facile recuperation, which is combined with an unaltered catalytic activity for all generations with respect to the monomeric catalyst. These observations arise from the overall molecular architecture of the dendronized polymer, which results in a stiffened, nanomanipulatable structure in which the dendrons are confined within a restricted space. The use of easily accessible hyperbranched polymers as economically viable substitutes for dendrimers used as catalyst supports may serve as the second example. In collaboration with Frey et al., dendritic catalyst were developed through both covalent and noncovalent anchoring of organometallic moi-

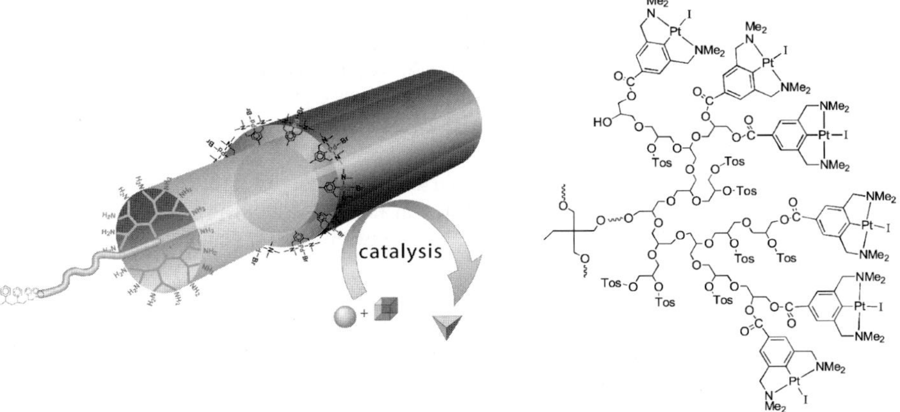

Fig. 11 Alternative dendritic macromolecular structures based on dendronized polystyrene (*left*) and hyperbranched polyglycidol (*right*) used as catalyst supports

eties to hyperbranched polyglycidol macromolecules (Fig. 11) [90, 91]. These metallodendritic species showed promising catalytic activities and recovery properties.

At the recycling level, however, the selection of the most appropriate catalyst and catalytic conditions (including the solvent), which should be compatible with the properties of the membrane, is a key issue. Although a large number of valuable membranes are commercially available, the lack of information on their compositions and chemical properties is a matter of compromise that complicates the full understanding of their interactions with the catalytic species and the reaction components.

In conclusion, the potential of soluble, nanosized metallodendrimers as catalysts in homogeneous reactions is well-consolidated. Future applications of these species are foreseen in high-tech nanotechnology applications in the fields of nano- and microreactors, cascade catalysis, and catalytic biomonitoring and biosensing. In this respect, the recent use of noncovalent strategies for the construction of multicomponent catalytic assemblies, and the use of biomacromolecules within dendritic structures is intriguing [60–62, 92, 93].

References

1. Newkome GR, He E, Moorefield CN (1999) Chem Rev 99:1689
2. Stoddart FJ, Welton T (1999) Polyhedron 18:3575
3. Niu Y, Crooks RM (2003) CR Chim 6:1049
4. Scott RWJ, Wilson OM, Crooks RM (2005) J Phys Chem B 109:692
5. Fréchet JMJ (2003) J Polym Sci Pol Chem 41:3713
6. Gillies ER, Fréchet JMJ (2005) Drug Discov Today 10:35

7. Astruc D (2004) J Organomet Chem 689:4332
8. Astruc D, Heuzé K, Gatard S, Méry D, Nlate S, Plault L (2005) Adv Synth Catal 347:329
9. Kreiter R, Kleij AW, Klein Gebbink RJM, van Koten G (2001) Top Curr Chem 217:163
10. Kleij AW, Klein Gebbink RJM, van Koten G (2002) Dendritic polymer application: catalysts. In: Fréchet J, Tomalia D (eds) Dendrimers and other dendritic polymers. Wiley, New York
11. Chase PA, Klein Gebbink RJM, van Koten G (2004) J Organomet Chem 689:4016
12. van de Coevering R, Klein Gebbink RJM, van Koten G (2005) Prog Polym Sci 30:474
13. Oosterom GE, Reek JNH, Kamer PCJ, van Leeuwen PWNM (2001) Angew Chem Int Ed 40:1828
14. Albrecht M, van Koten G (1999) Adv Mater 11:171
15. Albrecht M, Gossage RA, Lutz M, Spek AL, van Koten G (2000) Chem Eur J 6:1431
16. Albrecht M, Lutz M, Spek AL, van Koten G (2000) Nature 406:970
17. Albrecht M, Schlupp M, Bargon J, van Koten G (2001) Chem Commun 1874
18. Albrecht M, van Koten G (2001) Angew Chem Int Ed 40:3750
19. Dijkstra HP, van Klink GPM, van Koten G (2002) Acc Chem Res 35:798
20. van Heerbeek R, Kamer PCJ, van Leeuwen PWNM, Reek JNH (2002) Chem Rev 102:3717
21. Reek JNH, de Groot D, Oosterom GE, Kamer PCJ, van Leeuwen PWNM (2003) CR Chim 6:1061
22. Vankelecom IFJ (2002) Chem Rev 102:3779
23. Buhleier E, Wehner W, Vögtle F (1978) Synthesis 155
24. Tomalia DA, Baker H, Dewald J, Hall M, Kallos G, Martin S, Roeck J, Ryder J, Smith P (1985) Polym J 17:117
25. Newkome GR, Yao Z, Baker GR, Gupta VK (1985) J Org Chem 50:2003
26. Hawker CJ, Fréchet JMJ (1990) J Am Chem Soc 112:7638
27. Fréchet JMJ (1994) Science 263:1710
28. Achar S, Puddephatt RJ (1994) Angew Chem Int Ed 33:847
29. Achar S, Puddephatt RJ (1994) J Chem Soc Chem Commun 1895
30. Dahan A, Weissberg A, Portnoy M (2003) Chem Commun 1206
31. Dahan A, Portnoy M (2005) J Polym Sci Pol Chem 43:235
32. Dasgupta M, Peori MB, Kakkar AK (2002) Coord Chem Rev 233–234:223
33. Zhao M, Sun L, Crooks RM (1998) J Am Chem Soc 120:4877
34. Knapen JWJ, van der Made AW, de Wilde JC, van Leeuwen PWNM, Wijkens P, Grove DM, van Koten G (1994) Nature 372:659
35. Kleij AW, Gossage RA, Jastrzebski JTBH, Boersma J, van Koten G (2000) Angew Chem Int Ed 39:176
36. Kleij AW, Gossage RA, Klein Gebbink RJM, Brinkmann N, Reijerse, Kragl U, Lutz M, Spek AL, van Koten G (2000) J Am Chem Soc 122:12112
37. Kleij AW, Kleijn H, Jastrzebski JTBH, Smeets WJJ, Spek AL, van Koten G (1999) Organometallics 18:268
38. Kleij AW, Kleijn H, Jastrzebski JTBH, Spek AL, van Koten G (1999) Organometallics 18:277
39. Breinhauer R, Jacobsen EN (2000) Angew Chem Int Ed 39:3604
40. Hovestad NJ, Eggeling EB, Heidbüchel HJ, Jastrzebski JTBH, Kragl U, Keim W, Vogt D, van Koten G (1999) Angew Chem Int Ed 38:1655
41. Eggeling EB, Hovestad NJ, Jastrzebski JTBH, Vogt D, van Koten G (2000) J Org Chem 65:8857
42. Reetz MT, Lohmer G, Schwickardi R (1997) Angew Chem Int Ed 36:1526
43. Brinkmann N, Giebel D, Lohmer G, Reetz MT, Kragl U (1999) J Catal 183:163

44. de Groot D, Eggeling EB, de Wilde JC, Kooijman H, van Haaren RJ, van der Made AW, Spek AL, Vogt D, Reek JNH, Kamer PCJ, van Leeuwen PWNM (1999) Chem Commun 1623
45. de Groot D, Reek JNH, Kamer PCJ, van Leeuwen PWNM (2002) Eur J Org Chem 1085
46. de Groot D, Emmerink PG, Coucke C, Reek JNH, Kamer PCJ, van Leeuwen PWNM (2000) Inorg Chem Commun 3:711
47. Trinka TM, Grubbs RH (2001) Acc Chem Res 34:18
48. Schrock RR, Hoveyda AH (2003) Angew Chem Int Ed 42:4592
49. Connon SJ, Blechert S (2003) Angew Chem Int Ed 42:1900
50. Garber SB, Kingsbury JS, Gray BL, Hoveyda AH (2000) J Am Chem Soc 122:8168
51. Wijkens P, Jastrzebski JTBH, van der Schaaf PA, Kolly R, Hafner A, van Koten G (2000) Org Lett 2:1621
52. Togni A, Breutel C, Schnyder A, Spindler F, Landert H, Tijani A (1994) J Am Chem Soc 116:4062
53. Köllner C, Pugin B, Togni A (1998) J Am Chem Soc 120:10274
54. Schneider R, Köllner C, Weber I, Togni A (1999) J Chem Soc Chem Commun 2415
55. Togni A, Bieler N, Burckhardt U, Köllner C, Pioda G, Schneider R, Schyder A (1999) Pure Appl Chem 71:1531
56. Hearshaw MA, Moss JR (1999) J Chem Soc Chem Commun 1
57. Cuadrado I, Morán M, Casado CM, Alonzo B, Losada J (1999) Coord Chem Rev 395
58. Daniel M-C, Ruiz J, Astruc D (2003) J Am Chem Soc 125:1150
59. Valério C, Fillaut J-L, Ruiz J, Guittard J, Blais J-C, Astruc D (1997) J Am Chem Soc 119:2588
60. van de Coevering R, Kuil M, Klein Gebbink RJM, van Koten G (2002) J Chem Soc Chem Commun 1636
61. van de Coevering R, Alfers AP, Meeldijk JD, Martinez-Viviente E, Pregosin PS, Klein Gebbink RJM, van Koten G, submitted for publication to J Am Chem Soc
62. de Groot D, de Waal BFM, Reek JNH, Schenning APHJ, Kamer PCJ, Meijer EW, van Leeuwen PWNM (2001) J Am Chem Soc 123:8453
63. Bruner H (1995) J Organomet Chem 500:39
64. Peerlings HWI, Meijer EW (1997) Chem Eur J 3:1563
65. Seebach D, Rheiner PB, Greiveldinger G, Butz T, Sellner H (1998) Top Curr Chem 197:125
66. Ribourdouille Y, Engel GD, Gade LH (2003) CR Chim 6:1087
67. Oosterom GE, van Haaren RJ, Reek JNH, Kamer PCJ, van Leeuwen WNM (1999) Chem Commun 1119
68. Oosterom GE, Steffens S, Reek JNH, Kamer PCJ, van Leeuwen WNM (2002) Top Catal 19:61
69. Albrecht M, Hovestad NJ, Boersma J, van Koten G (2001) Chem Eur J 7:1289
70. Arink AM, van de Coevering R, Wieczorek B, Firet J, Jastrzebski JTBH, Klein Gebbink RJM, van Koten G (2004) J Organomet Chem 689:3813
71. Arink AM, Braam TW, Keeris R, Jastrzebski JTBH, Benhaim C, Rosset S, Alexakis A, van Koten G (2004) Org Lett 6:1959
72. Haubrich A, van Klaveren M, van Koten G, Handke G, Krause N (1993) J Org Chem 58:5849
73. Spescha M, Rihs G (1993) Helv Chim Acta 76:1219
74. van Klaveren M, Lambert F, Eijkelkamp JFM, Grove DM, van Koten G (1994) Tetrahedron Lett 35:6135
75. Zhou QL, Pfaltz A (1994) Tetrahedron 50:4467
76. Seebach D, Jaeschke G, Pichota A, Audergon L (1997) Helv Chim Acta 80:2515

77. Jansen MD, Grove DM, van Koten G (1997) Prog Inorg Chem 46:97
78. Knotter DM, van Koten G, van Maanen HL, Grove DM, Spek AL (1989) Angew Chem Int Ed 28:341
79. Knotter DM, Janssen MD, Grove DM, Smeets WJJ, Horn E, Spek AL, van Koten G (1991) Inorg Chem 30:4361
80. Knotter DM, Grove DM, Smeets WJJ, Spek AL, van Koten G (1992) J Am Chem Soc 114:3400
81. Kleijn H, Rijnberg E, Jastrzebski JTBH, van Koten G (1999) Org Lett 1:853
82. Gossage RA, Jastrzebski JTBH, van Koten G (2005) Angew Chem Int Ed 44:1448
83. Dijkstra HP, Steenwinkel P, Grove DM, Lutz M, Spek AL, van Koten G (1999) Angew Chem Int Ed 38:2185
84. Dijkstra HP, Meijer MD, Patel J, Kreiter R, van Klink GPM, Lutz M, Spek AL, Canty AJ, van Koten G (2001) Organometallics 20:3159
85. Dijkstra HP, Kruithof CA, Ronde N, van de Coevering R, Ramón DJ, Vogt D, van Klink GPM, van Koten G (2003) J Org Chem 68:675
86. van Klink GPM, Dijkstra HP, van Koten G (2003) CR Chim 6:1079
87. Dijkstra HP, Ronde N, van Klink GPM, Vogt D, van Koten G (2003) Adv Synth Cat 345:364
88. Dijkstra HP, Albrecht M, Medici S, van Klink GPM, van Koten G (2002) Adv Synth Cat 344:1135
89. Suijkerbuijk BMJM, Shu L, Klein Gebbink RJM, Schlüter AD, van Koten G (2003) Organometallics 22:4175
90. Stiriba S-E, Slagt MQ, Kautz H, Klein Gebbink RJM, Thomann R, Frey H, van Koten G (2004) Chem Eur J 10:1267
91. Slagt MQ, Stiriba S-E, Klein Gebbink RJM, Kautz H, Frey H, van Koten G (2002) Macromolecules 35:5734
92. Volkmer D, Bredenkötter B, Tellenbröker J, Kögeler P, Kurth DG, Lehmann P, Schnablegger H, Schwan D, Piepenbrink M, Krebs B (2002) J Am Chem Soc 124:10489
93. van Baal I, Malda H, Synowsky SA, van Dongen JLJ, Hackeng TM, Merkx M, Meijer EW (2005) Angew Chem Int Ed 44:5052

Top Organomet Chem (2006) 20: 39–59
DOI 10.1007/3418_031
© Springer-Verlag Berlin Heidelberg 2006
Published online: 23 June 2006

Supramolecular Dendritic Catalysis:
Noncovalent Catalyst Anchoring
to Functionalized Dendrimers

Fabrizio Ribaudo · Piet W. N. M. van Leeuwen · Joost N. H. Reek (✉)

Homogeneous and Supramolecular Catalysis, van't Hoff Institute for Molecular Sciences,
University of Amsterdam, Nieuwe Achtergracht 166, 1018 WV Amsterdam,
The Netherlands
reek@science.uva.nl

Abstract The development of catalytic systems that provide rapid and selective chemical transformations and can be completely separated form the product is still a major challenge. Dendrimers can be used as well-defined nanosized supports for the immobilization of homogeneous catalysts leading to homogeneous catalysts that can be easily removed from the reaction mixture by nanofiltration. So far most effort has been put into the covalent functionalization of dendritic structures with catalytic sites. A recent novel strategy in this research area comprises the noncovalent anchoring of catalysts to support applying well-defined binding sites, offering several advantages above the traditional covalent approaches. In this chapter we will discuss the supramolecular approach in detail and we will highlight the progress in the area of supramolecular dendritic catalysis.

Keywords Dendrimers · Transition metal catalysis · Supramolecular chemistry · Catalyst recovery

1
Introduction

Dendrimers are well-defined hyperbranched macromolecules that have become particularly popular in the preceding decades. Rapid developments of synthetic procedures have made functionalized dendrimers readily available in sufficient quantities to facilitate the rapid development of dendrimer chemistry. Despite the huge effort of many groups, most synthetic procedures result in dendrimers that have defects, especially for the larger generation, but the consequences for their properties are generally thought to be small. Several potential applications for dendrimers, including catalysis [1–8], are well documented in numerous reviews [9–12].

In the early 1990s, dendrimers were used for the first time as support for catalysis [13], aiming at an optimum combination of homogeneous catalysis and easy separation from reaction products by modern membrane techniques. As is clear from the reviews on dendritic catalysis [1–8] and the previous section, there are numerous examples of dendritic transition metal catalysts where the active center is attached to various dendritic backbones at different locations. It was proposed that dendrimers potentially could modify the catalyst performance; such dendritic effects in catalysis include increased or decreased activity, selectivity and stability. Especially dendrimers functionalized at their cores offer the prospect of mimicking the properties of enzymes, which can be regarded as their natural counterparts, and such dendrimers may yield catalysts with unique activities or stabilities. In core- (and focal point-) functionalized dendrimers, the catalyst may benefit from the site isolation created by the environment of the dendritic structure. When reactions are deactivated by excess ligand or when a bimetallic deactivation mechanism is operative, core-functionalized dendrimers can minimize the deactivation. Periphery-functionalized dendrimers contain multiple reaction sites and may have extremely high local catalyst concentrations, which can lead to cooperative effects in reactions that proceed via a bimetallic mechanism [14, 15]. This may be a potential detriment of periphery-functionalized dendrimers, because several deactivation pathways operate via bimetallic mechanisms (e.g., ruthenium-catalyzed metathesis [16], palladium-catalyzed reductive coupling of benzene and chlorobenzene [17], and reactions involving radicals [18]).

Besides the hitherto described dendritic effects, the main aspect of interest regarding the use of dendritic supports in catalysis is represented by the possibility of recovering the catalyst. It is clear from the contributions of many research groups that dendrimers are suitable supports for recyclable transition metal catalysts. Separation and/or recycle of the catalysts are possible with these functionalized dendrimers; for example, separation results from precipitation of the dendrimer from the product liquid; two-phase catalysis allows separation and recycling of the catalyst when the products and catalyst are concentrated in two immiscible liquid phases; and immobilization of the dendrimer in an insoluble support (such as cross-linked polystyrene or silica) allows use of a fixed-bed reactor holding the catalyst and excluding it from the product stream. For dendritic catalysts separation with these traditional techniques, the function of the dendrimer is not always clear. In contrast, the large

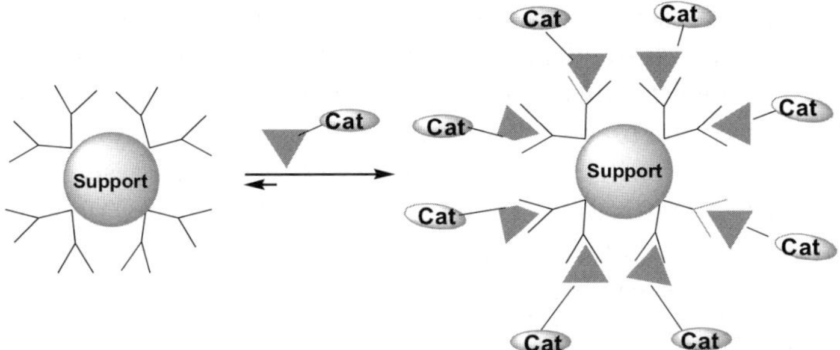

Fig. 1 Schematic presentation of supramolecular anchoring of catalysts to dendritic support

size and the globular structure of the dendrimers enable efficient separation by nanofiltration techniques. Nanofiltration can be performed either batchwise or in a continuous-flow membrane reactor (CFMR), which offers major advantages, particularly for reactions that benefit from low substrate (reactant) concentrations or suffer from side reactions of the products. Membrane reactors have been investigated since the 1970s [19]. Although membranes can have several functions in a reactor, the most obvious is the separation of reaction components. Initially the focus has been mainly on polymeric membranes applied in enzymatic reactions, and ultrafiltration of enzymes is commercially applied on large scale for the synthesis of fine chemicals (e.g., L-methionine) [20–24]. Membrane materials have been improved significantly over those applied initially, and nanofiltration membranes suitable to retain relatively small compounds are now available commercially (e.g., mass cut-off of 400–750 D).

Whether dendritic catalysis can compete successfully in commercial applications with other strategies that allow recycling of the catalyst remains to be seen. Dendrimer supports are still relatively expensive, but for applications that do not require the well-defined structure of the dendrimer support, hyperbranched polymers can offer a cheaper alternative. A recent novel strategy in this research area that might provide the added value required to make the step towards commercial applications involves the noncovalent functionalization of dendritic support with catalysts (Fig. 1). This offers several advantages above the traditional covalent approaches. In this chapter we will review the progress in the area of supramolecular dendritic catalysis.

2
Covalent Versus Noncovalent Anchoring of Catalysts to Dendrimers

Before describing examples of supramolecular dendritic catalyst we will first evaluate the potential advantages and disadvantages of covalent and noncovalent functionalization of dendrimers. The covalent functionalization of dendrimers with catalytic sites requires the development of novel synthetic routes for each novel catalyst system. This is very labor-intensive and therefore costly. Supramolecular anchoring involves only one synthetic route that results in dendrimers with well-defined binding motifs. Catalysts only need to be functionalized with the complementary motif, which is generally much easier than multiple-functionalization of the dendritic framework. Thus, one type of dendritic support can be used to support many different catalyst systems. In addition, the reversible nature of the noncovalent approach allows controlled de- and re-functionalization of the support, which enables the easy reuse of the homogeneous catalyst *and* the dendritic support. It also gives rise to new opportunities such as easy variation of catalyst loading, even during catalysis, and replacing decomposed catalyst by new material. Indeed, when the catalyst that is covalently attached to the dendrimer has decomposed, both the catalyst and the dendritic support will be wasted. In general the supramolecular approach is far more flexible and general because of its inherent modular nature. Of course the supramolecular approach also has its disadvantages, or research challenges if you like. First of all, the supramolecular binding event should be compatible with the catalytic event. This means that the supramolecular strategy poses restriction on the conditions under which the dendritic catalyst can be applied. For example, if the binding is based on hydrogen bonding, apolar organic solvents should be used. Also, many transition metal-catalyzed reactions require basic conditions, and the binding motifs that are used should be base-stable in these cases.

Another important issue evolves if the dendritic catalyst will be used in a continuous-flow membrane reactor. Generally, two forms of leaching have to be considered when dendritic transition metal catalysts are used in such reactors: depletion of the dendritic catalyst through the membrane and metal dissociation (possibly after decomposition) from the dendrimer resulting

Table 1 Advantages and disadvantages of covalently and noncovalent anchoring of catalysts to dendrimers

	Covalent attachment	Noncovalent attachment
Multipurpose use of dendritic scaffold	No	Yes
Multiple use of dendritic scaffold	No	Yes
Easy variation of catalyst loading on support	No	Yes
Catalyst recycling via nanofiltration	Yes	Depends on association constant

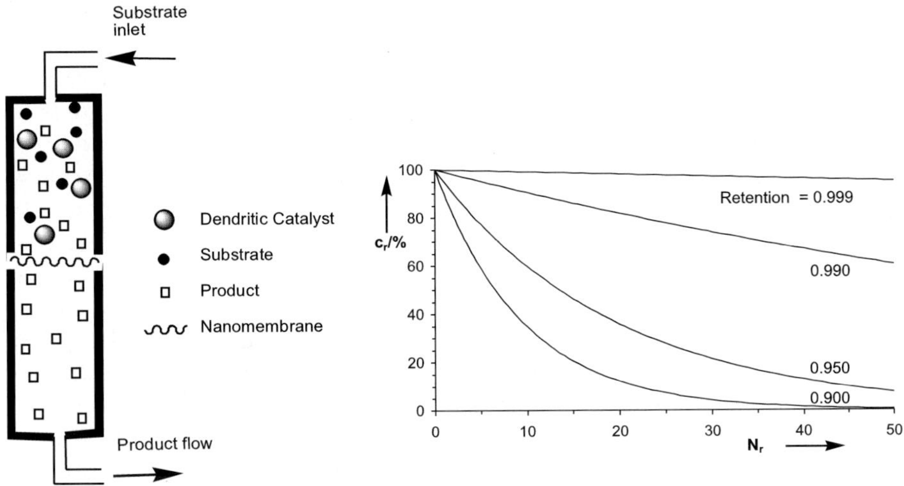

Fig. 2 Schematic presentation of a membrane reactor (*left*) and theoretical relative concentrations (Cr) of the dendritic species versus the substrate flow (in residence times Nr) calculated for various retention factors

in further leaching of the unsupported metal through the membrane. In addition to these leaching issues, the supramolecular dendritic catalyst has a third pathway that leads to catalyst depletion, namely dissociation of the catalyst system from the binding site. Clearly, this process is strongly dependent on the association constant (binding strength of the motif used) and the absolute concentration of the dendrimer (with the binding sites) and the catalyst. The question that rises is how strong the binding of the catalyst to the dendritic support should be. In Fig. 2 the theoretical retention of catalysts are plotted for dendritic systems with various retention factors[1]. For example, if a dendritic catalyst has a retention factor of 0.95, only 25% of the catalyst would remain in the reactor after the reactor had been flushed with 30 times its volume. For practical applications, the overall retention of the dendritic catalyst must be extremely high (typically, > 99.9%) to maintain the material in a CFMR for long reaction times. The required retention obviously depends on the application; processes for the commodity chemical industry generally require higher numbers of turnovers and therefore more efficient catalyst recycling than those for high added-value fine chemicals. In Fig. 3 we have plotted theoretical curves with retention of catalysts systems that are anchored to dendrimers with various association constants. We have assumed for this particular example that the retention of the dendrimer is high (100%)

---

[1] **Retention factors** were calculated using equation: retention factor $= 1 + \{\ln[a/(a+b)]/x\}$, where a is the amount of dendrimer inside the reactor after the experiment; b is the amount of dendrimer that went through the membrane; and x is the number of reactor volumes flushed with substrate solution.

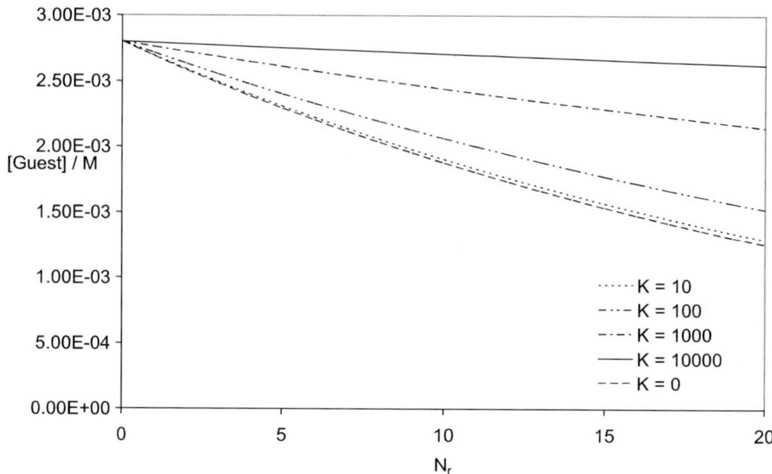

Fig. 3 Theoretical retention curves for catalysts noncovalently anchored to a function-alized dendritic host with various association constants: concentration guest catalyst ([Guest]) in the reactor as function of the residence time (Nr). Calculation parameters: [Host] = 3.60×10^{-3} M, [Guest]start = 2.80×10^{-3} M, retention guest catalyst = 96%, re-tention host = 100%

and that of the free catalyst (not associated to the dendrimer) is also rather high (96%). The retention of the catalyst is typical what we find if we use membranes with a cut-off around 700 D. From the theoretical curves it can be seen that the association constant of the catalyst to the dendrimer must be typically above 10^4 M^{-1} to obtain catalyst systems that are well suited for ap-plication in continuous-flow membrane reactors. If membranes are used that have a larger cut-off, between 1000 and 2000, the retention of the free cata-lyst will drop dramatically in which case a high association constant becomes even more important.

3
Noncovalently Functionalized Dendrimers Based on Multiple Interactions

It is interesting to note that supramolecular chemistry using dendritic scaffold has been explored since many years [25]. The first example in which a supramolecular dendritic catalyst is described was reported only recently [26]. Phosphine ligands were anchored in the periphery of poly(pro-pylene imine) dendrimers using a specific binding motif that is complemen-tary to that of the dendrimer support. The dendrimer provided directional binding sites for the strong but reversible binding of 32 guest molecules func-tionalized with the complementary binding motif. The binding is based on a combination of multiple hydrogen bonds and ionic interactions (Fig. 4).

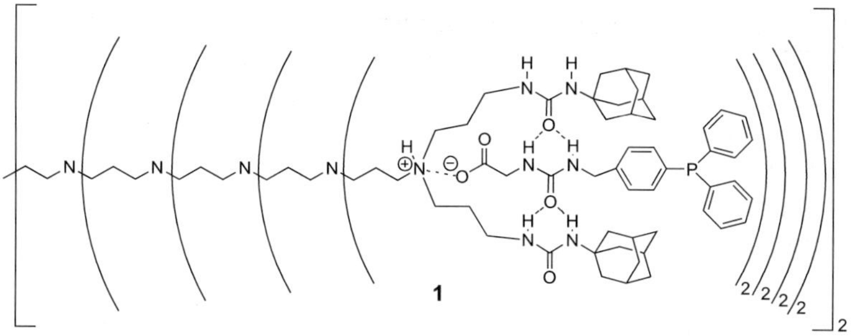

Fig. 4 The first supramolecular dendritic catalyst in which 32 phosphine ligands are noncovalently anchored to the functionalized dendrimer by hydrogen bonds and ionic interactions

The binding constant of the anchor into the periphery of the dendrimer and that of the palladium to the ligand are very high; the guest-Pd-dendrimer complex remained intact during size exclusion chromatography (SEC). The dendrimer containing 32 phosphine-functionalized guest molecules was used as a multidentate ligand in the palladium-catalyzed allylic amination with crotyl acetate and piperidine as the reactants. The reaction was fast; with a substrate to Pd ratio of 30, the conversion was more than 80% after 10 min. Approximately the same rate and selectivity (branched : *trans* : *cis ratio* = 61 : 33 : 6) were observed for the monomeric palladium complex in absence of the dendrimer.

The supramolecular guest-Pd-dendrimer complex was found to have a retention of 99.4% in a continuous-flow membrane reactor. This indicates that the supramolecular anchoring is sufficiently strong in this example for application in such process. The supramolecular system was also investigated as a catalyst for the allylic amination reaction. A solution of crotyl acetate and piperidine in dichloromethane was pumped through the reactor. The conversion reached its maximum (ca. 80%) after approximately 1.5 h (which is equivalent to 2–3 reactor volumes of substrate solution pumped through the reactor). The conversion remained fairly constant during the course of the experiment (Fig. 5). The red dot indicates the expected conversion (on the basis of the retention time) after the amount of 13 reactor volumes of substrate solution had been pumped through the reactor. The small decrease in conversion observed was attributed to deactivation of a part of the catalyst, which was also observed for covalently functionalized systems [26]. These initial experiments clearly demonstrate that noncovalently functionalized dendrimers are suitable as soluble and recyclable supports for catalysts provided that the binding constant is sufficiently high.

After these initial experiments, carbosilane dendrimers with a similar binding motif in the core have been synthesized and their properties have been investigated [27] (van Heerbeek et al., private communications). It was found that guest ligand molecules (L in Table 2) indeed were bound in the bis-urea binding pocket in the core of the dendrimer with association constants around $10^4\,M^{-1}$. Reactions carried out in batch processes showed that if the catalyst was anchored in the core of larger dendrimers, the reaction was slower than that of the free catalyst. A similar observation was observed with covalently core-functionalized systems [28], indicating that the reaction indeed takes place in the core of the dendrimer. Also in the noncovalently

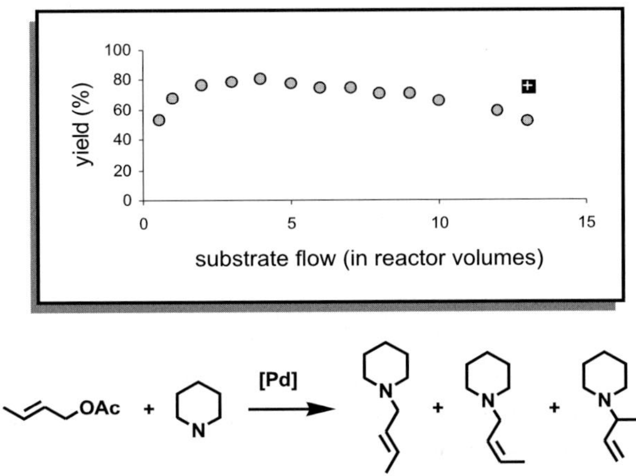

Fig. 5 The application of a noncovalently functionalized dendrimer (1) in a CFMR in the allylic amination of crotyl acetate and piperidine in dichloromethane (Koch MPF-60 NF membrane, molecular weight cut-off = 400 Da). The *crossed square* represents the yield expected on the basis of the retention factor of the system and the deviation is explained by catalyst decomposition

Table 2 Palladium-catalyzed allylic amination using catalysts that are noncovalently anchored in the core of a carbosilane dendrimer

Catalyst	Conversion	N–<	N–/\	N–\/
L$_2$Pd(crotyl)Cl	92	61%	31%	7%
G3{L$_2$Pd(crotyl)Cl}	85	61%	31%	7%
G4{L$_2$Pd(crotyl)Cl}	61	60%	32%	8%

Conditions: [Pd] = 0.01 mM; [crotylacetate] = 1 mM; [piperidine] = 2 mM; Gx : L : Pd = 1.3 : 2.2 : 1; solvent CH$_2$Cl$_2$; r.t.

core-functionalized systems the catalyst activity is reduced because the catalytic center is less accessible for the substrates. Preliminary experiments show that the core-functionalized system can also be applied in a continuous-flow membrane reactor (Fig. 6).

Recently, we extended the approach to dendrimers immobilized on silica. For this purpose, the first generation (and also larger generations) DAB dendrimer was functionalized with urea propyl-(SiOR)$_3$ groups that via an easy process were grafted on silica. The catalyst was noncovalently anchored by simply adding a solution of a metal complex with ancillary ligands utilized with the typical the complementary motif (Fig. 7) [29].

The supramolecular interaction between the transition metal catalyst and the binding site is sufficiently strong to enable efficient catalyst recycling. In addition, the support can be readily re-functionalized with different catalyst systems, by washing with methanol to remove the first catalyst system and then attaching the new catalyst system by just stirring in apolar solvent such as toluene (Fig. 7).

The resulting noncovalently immobilized complexes have been used as ligand systems for both the Pd-catalyzed allylic amination reaction and the Rh-catalyzed hydroformylation. A glycine-urea functionalized PPh$_3$ ligand, 4(S), was noncovalently attached to the immobilized dendritic support, and the application of this system in the Pd-catalyzed allylic amination attains similar yields and product distributions as the homogeneous analogue for the

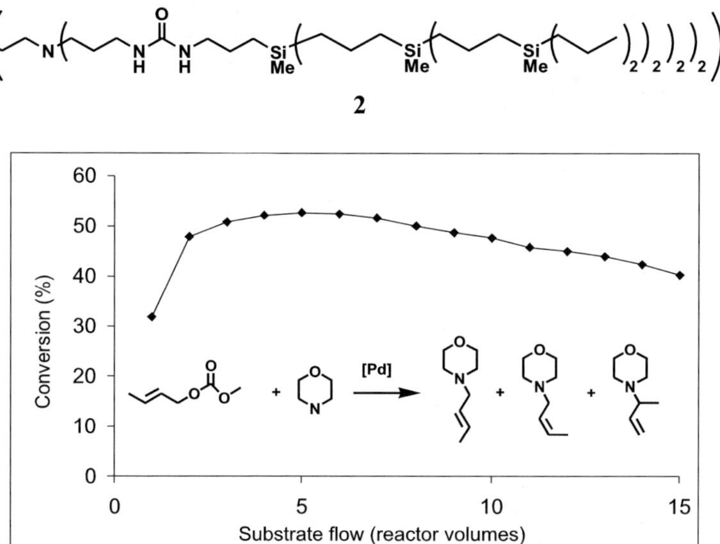

Fig. 6 The application of a noncovalently core-functionalized dendrimer (2) in a CFMR in the allylic amination of crotyl carbonate in dichloromethane (Koch MPF-60 NF membrane, molecular weight cut-off = 400 Da)

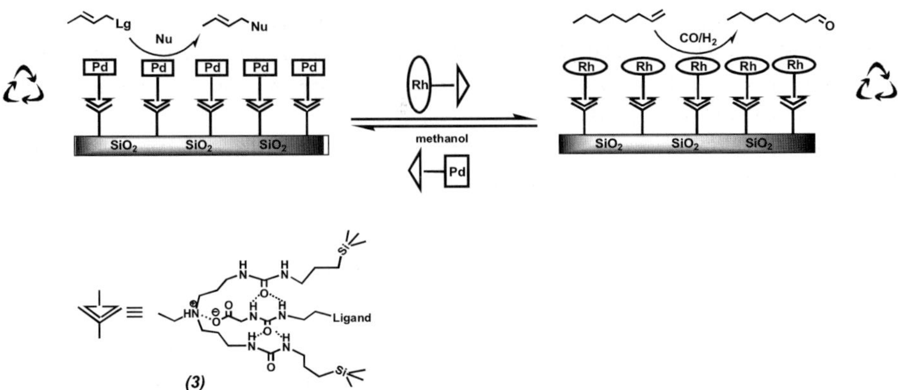

Fig. 7 The concept of supramolecular anchoring of catalysts to dendritic binding sites immobilized on silica (A) and the binding motif based on hydrogen bonding (HB) that has been used

Pd-catalyzed allylic amination of crotyl acetate by piperidine, while exhibiting a reduced rate as is commonly observed for heterogenized systems (90% conversion achieved after 30 min compared with 5 min for the homogeneous system). Interestingly, the catalyst could be recycled three times via a simple filtration step. Subsequently, the catalyst was separated from the support and the support was uploaded with hydroformylation catalysts.

Ligand **4** was first studied in the rhodium-catalyzed hydroformylation of 1-octene and gave chemo- and regio-selectivities typical of bis-triphenylphosphine based rhodium complexes. Reactions proceeded to 80–90% conversion, and as usual for supported catalysts the activity was less than the analogous homogeneous systems. The catalyst can be recycled up to eight times with only a slight drop in activity, which is due to metal-leaching. A rhodium catalyst based on the glycine-urea functionalized Xantphos ligand

(5) was subsequently used in combination with the same support material, as a catalyst for the hydroformylation of 1-octene. In 11 consecutive reactions the catalyst did not show any deterioration or metal-leaching. Similar to what was previously observed for covalently anchored systems (6), a decrease in activity and selectivity is observed compared to the homogeneous system. Interestingly, higher activity and selectivity for the linear aldehyde are observed for the noncovalently anchored ligand compared to covalently anchored Nixantphos (6), while in the homogeneous phase these Nixantphos and Xantphos ligands show similar activity and selectivity [30, 31].

These noncovalently anchored catalysts in general exhibit a behavior similar to their covalently bound analogues, but can now be separated from the support after the reactions by a simple filtration step. So far, these immobilized systems have not been used in continuous flow reactors.

4
Noncovalently Functionalized Dendrimers
Based on Single Point Interactions

Van de Coevering et al. [32] utilized ion-pairing interactions to tether transition metal complexes with sulfonated anionic tails to cationic dendrimeric supports (Fig. 8). In these assemblies, eight anionic organometallic Pd-

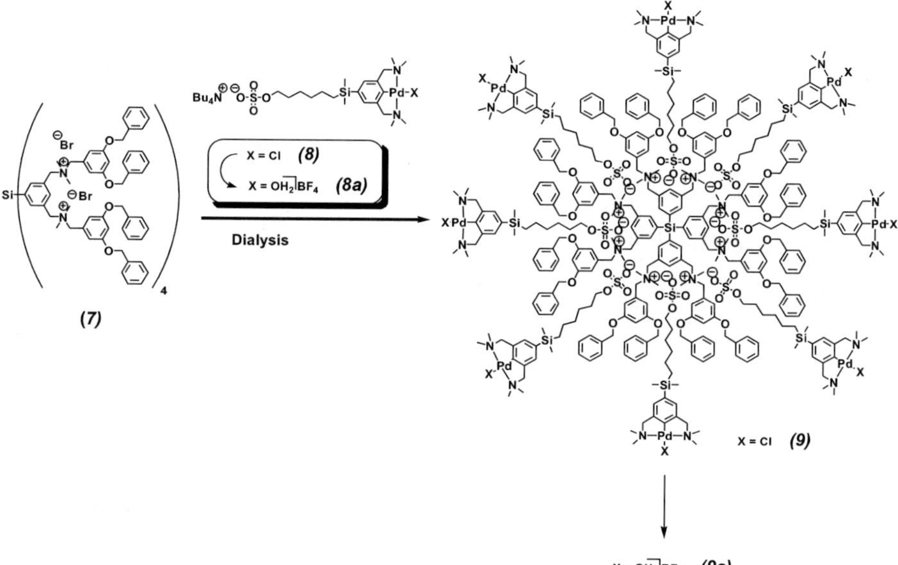

Fig. 8 Stoichiometric assembly of an octa-cationic core-shell dendrimer and eight anionic Pd(II) guest molecules

fragments are incorporated by the cationic core-shell dendritic structure, i.e., the number of incorporated mono-anionic Pd(II)-guests is determined by the number of cationic sites in the dendrimer core.

According to the NMR studies, the organometallic Pd(II) moieties are located at or around the dendrimer periphery depending on the tether size and the dendrimer generation. The diffusion coefficients of the octa-cation and the eight mono-anions in these dendritic assemblies show a good correlation and indicate an enhanced "entrapment" of the anions by the larger dendrimers. These assemblies were successfully applied as Lewis acid catalysts in the cyclization of benzaldehyde and methyl isocyanate. In all cases, the conversions after 24 h were somewhat lower than those of the unsupported analogues, whereas the product *trans/cis* ratio was retained in the presence of the dendritic support. These observations suggest that each supported Pd(II) moiety in the dendritic assembly behaves kinetically independently and that, apparently, the "embedding" in the dendrimer framework does not produce mass transfer limitations phenomena that would alter the reaction kinetics. No experiments of recovery and recycling were reported.

Kaneda et al. [33, 34] have recently utilized acid-base interactions to non-covalently attach diphenylphosphine-4-benzoic acid palladium complexes to the exterior of poly(propyleneimine) dendrimers and to cavities within these dendrimers (Fig. 9) via the interaction of the benzoic acid with the carboxyl groups of amide moieties positioned within the dendrimer. Interestingly, the supramolecular dendritic catalyst gave rise to much more active catalysts for

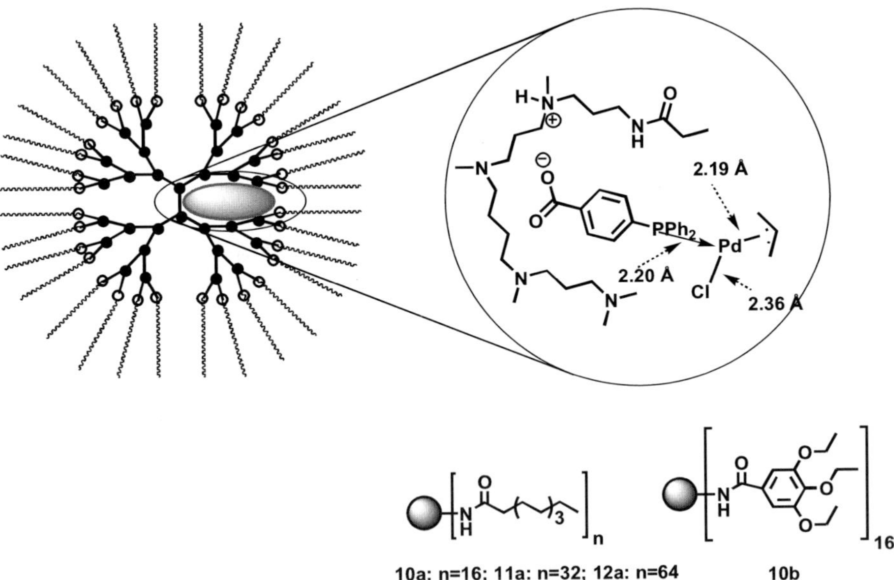

Fig. 9 Proposed structure of dendrimer-encapsulated Pd complexes

the Heck reaction. In addition, selectivity for mono-substitution in the Heck reaction of *p*-diiodobenzene with *n*-butyl acrylate was observed for the dendritic encapsulated system, whereas the parent complex gave rise to a mixture of mono- and di-substituted products. The origin of this dendritic selectivity effect remains unclear. In the allylic amination of cinnamyl methyl carbonate, using morpholine as the nucleophile, the supramolecular catalysts also gave rise to higher selectivities compared to the non supported analogue. Functionalization of the exterior of the dendritic scaffold with triethoxybenzoyl chloride instead of decanoyl chloride makes the dendrimer insoluble in aliphatic solvents, but soluble in polar solvents such as DMF. These properties enabled a simple, recyclable thermomorphic biphasic system for the allylic amination to be devised.

Polyoxometallate clusters (POM's), a class of well-defined structures that are rich in redox chemistry and therefore highly suitable for oxidation catalysis have been covalently attached [35] to dendritic support to aid separation, recycling and stability. The resulting tetra-(POM) molecules are hydrolytically stable, catalyze peroxide oxidations and represent the initial examples of a potentially very large class of new POM-containing catalytic materials. Very recently Astruc et al. [36] noncovalently linked POM's to a dendritic support by utilizing ion-pairing interactions between the anionic POM fragments and a cationic dendrimer. The polyoxometallate fragment $[PO_4\{WO(O_2)_2\}_4]^{3-}$ (Fig. 10) was noncovalently linked to tri-cationic tripod fragments of the dendrimer yielding an epoxidation catalyst that was as active as its unbound analogue. A comparison of the homogeneous mono-, bis-, tris-, and tetra(POM) catalysts indicated that there was no measurable dendritic effect on the reac-

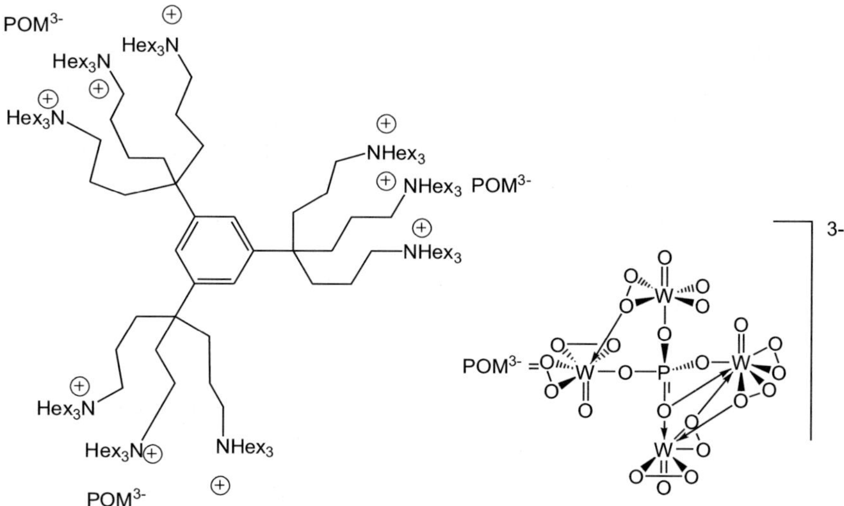

Fig. 10 Polyoxometallate clusters anchored to soluble support *via* ionic interactions

tion kinetics within this series. A dendritic effect, however, was noted in the recovery of the catalysts. They were recovered by precipitation and filtration with a recovery efficiency of up to 96%. These catalysts could then be re-used with no loss of activity.

5
Noncovalently Functionalized Hyperbranched Polymers

The structural perfection of a dendritic support is not required for every application in catalysis, and hyperbranched polymers provide interesting and cheap alternatives as catalyst supports [37]. These hyperbranched polymers are obtained from a simple one-pot synthesis, yielding globular polymeric structures with broad weight distributions compared to their dendritic analogues.

Mecking [38] utilized electrostatic interactions to bind phosphine ligands with multiple sulfonate groups, such as NaTPPTS (tris(sodium-m-sulfonatophenyl)phosphine), to soluble linear polyelectrolytes which can be recovered and recycled by ultrafiltration. The electrostatically polymer-bound complex was employed in the hydroformylation of 1-hexene, yielding turn-over-frequencies of up to 160 TOh^{-1} at 80 °C and 30 bar (CO/H$_2$ = 1). The system exhibits the typical selectivity ($l : b$ = 2.5–3.1) of a *bis*(triphenylphosphine) bound rhodium catalyst. This approach has recently been extended to hyperbranched supports [39]: polyelectrolytes with Ph$_2$P(C$_6$H$_{4-}$$p$-SO$_3^-$) counterions were prepared by ion exchange from hyperbranched polycations with a polyglycerol-based polyether scaffold and 1,2-dimethylimidazolium end groups. Clear advantages of highly branched macromolecules by comparison to linear polymers are their low tendency for crystallization and corresponding high solubility, which is beneficial in catalyst synthesis and enables high conversions. Hydroformylations were carried out in homogeneous solution using the aforementioned polyelectrolyte as polymer-bound ligand and [Rh(CO)$_2$(acac)] as a metal precursor. In the hydroformylation of 1-hexene, a slightly reduced catalytic activity of the polymer-bound catalyst but somewhat higher linear/branched selectivities compared to the unbound complex were observed, possibly caused by the high local phosphine concentration at the polymer periphery. The noncovalently bound catalyst could be recovered and recycled by ultrafiltration several times, but the system was not tested in a closed, continuously operated membrane reactor.

Frey and Van Koten et al. [40–42] reported on the noncovalent encapsulation of sulfonated pincer-platinum(II) complexes in readily available amphiphilic nanocapsules based on hyperbranched polyglycerol, possessing a reverse micelle-type architecture. The incorporated platinum(II) complexes showed catalytic activity in a double Michael addition, albeit with decreased activities compared to the free pincer complex. Due to the size of

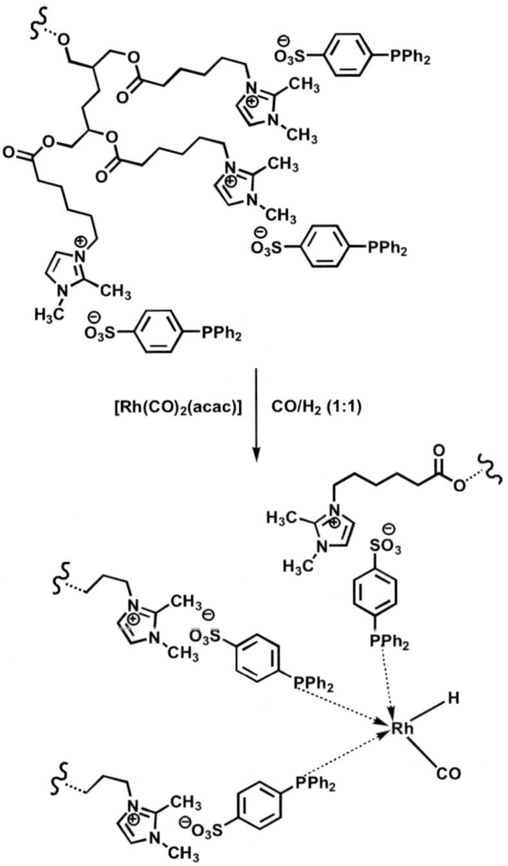

[Rh(CO)$_2$(acac)] | CO/H$_2$ (1:1)

Fig. 11 Schematic representation for the in situ formation of the catalyst precursor with polymeric ligands of the type PG(C$_5$-1,2DMI-TPPMS)$_{1.0}$

the nanocapsules, it was possible to separate the products from the encapsulated catalysts by dialysis, and recover > 97% of the catalytic material. It is unknown if the 3% loss is due to the supramolecular nature of the system. The same groups have synthesized optically active hyperbranched polymers and used them to immobilize platinum NCN-pincer complexes both covalently and noncovalently. These chiral supports exhibit a slight circular dichroic activity, but no enantioselectivity is observed upon using these systems to catalyze Michael additions between methyl vinyl ketone and ethyl α-cyanopropionate. Quantitative recovery of these catalysts from catalytic reactions was achieved by dialysis.

Haag et al. [43] have recently reported on the supramolecular immobilization of a perfluoro-tagged Pd catalyst on a dendritic polyglycerol ester with a perfluorinated shell and they have investigated its activity in Suzuki coup-

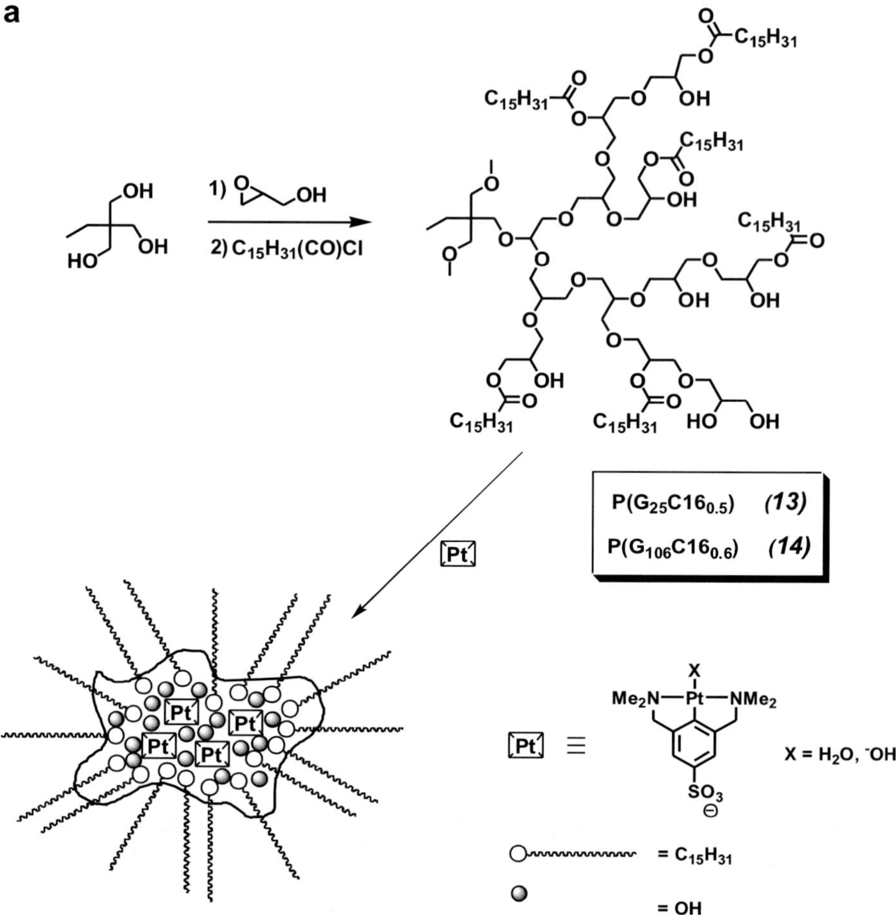

Fig. 12 a Molecular nanocapsule synthesis, structure, and noncovalent encapsulation of platinum pincer complexes in the hydrophilic interior

ling reactions (in DMF as solvent). The strong interaction between the perfluorinated chains in the supramolecular complex was studied by ^{19}F NMR spectroscopy. The spectrum shows an isolated CF_3 signal at -83.31 ppm. Upon complex formation a significant amount of the CF_3 signal undergoes a down-field shift of 1.70 ppm to -81.61 ppm. This shift of the CF_3 signal is characteristic of the partial intercalation of perfluoroalkyl chains. Concerning the catalytic activity of the immobilized catalyst it is noteworthy that the fluorous–fluorous interaction is temperature dependent and is reduced significantly at higher temperatures. This will enhance the flexibility of the ligand, which might increase its catalytic activity. The performances of the immobilized perfluoro-tagged catalyst was studied in homogeneous

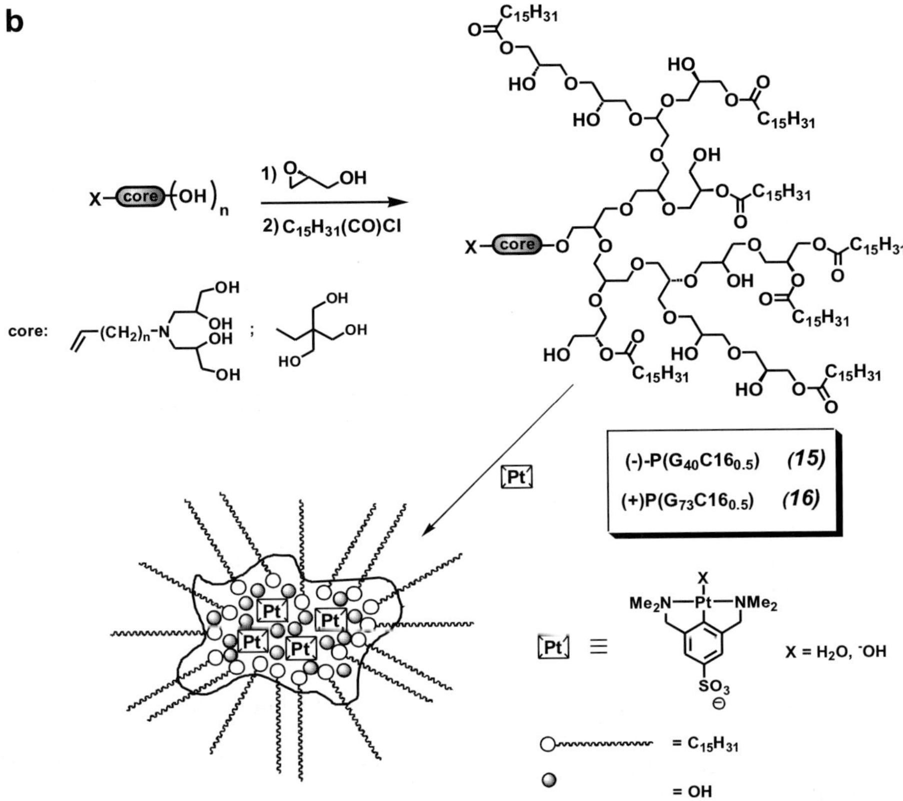

Fig. 12 b Preparation of chiral molecular nanocapsules 15 and 16 with Bis(2,3-dihydroxypropyl)-10-undecenylamine and trimethylolpropane (TMP), Respectively, and a schematic encapsulation of sulfonated platinum pincer complexes of type 3 in the hydrophilic compartment of the nanocapsules

Fig. 13 a Synthesis of dendritic polyglycerol perfluorolkyl ester 18 from hyperbranched polyglycerol 17

Suzuki coupling reactions (solvent DMF, temp. 80 °C) and compared to the unsupported perfluoro-tagged catalyst at different catalyst concentrations. The yields using the perfluoro-tagged catalyst are significantly lower compared to the reactions were the supramolecular dendritic catalyst assembly

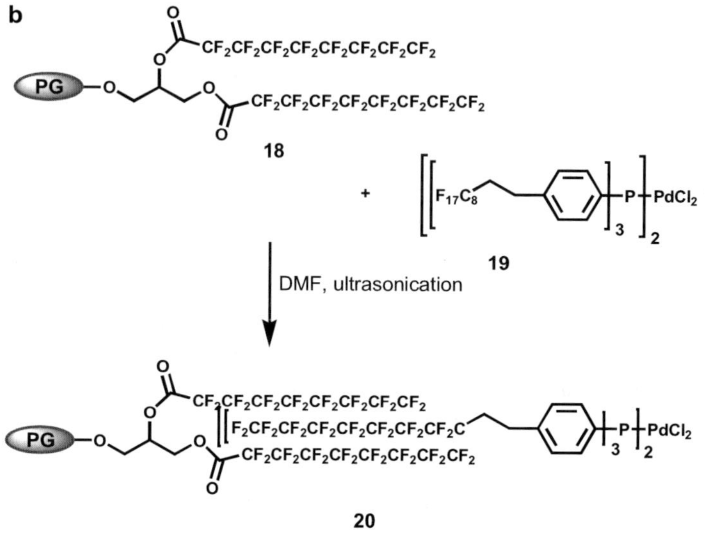

Fig. 13 **b** Supramolecular interaction (20) between perfluoro-tagged Pd complex (19) and polyglycerol ester 18

was applied. This is probably due to the fact that the "bare" perfluoro-tagged Pd catalyst is not soluble in DMF, in contrast to the supramolecular dendritic catalyst assembly. The supramolecular supported catalyst could be separated from the products by simple precipitation using a mixture of DME/water for recycling and multiple use purposes.

6
Conclusions and Outlook

In this contribution, the noncovalent immobilization of homogeneous catalysts to dendritic and other soluble macromolecular supports has been surveyed. In the examples reported so far, the supramolecular noncovalent immobilization of a homogeneous catalyst is based various interactions. In the first examples, the anchoring process was based on multiple interactions between catalyst and dendritic support, providing well-defined nano-sized catalyst complexes. It has been shown that functionalization of the dendritic scaffold can also rely on simple ion exchange or acid-base ion-pairing. It is clear that dendritic scaffolds are suitable supports to prepare supported transition metal catalysts using supramolecular strategies, and several transition metal-catalyzed reactions have been shown to be compatible with the noncovalent anchoring process. This requires that the functional groups used for the anchoring and the catalysts, substrates and products are compatible, and that reaction conditions are such that the catalysis proceeds and the binding is possible. Although it has been proven, much more studies are required to learn more in detail about these aspects since in many examples it is not exactly know how, for example, the formation of the product affects the binding process of the catalysts to support.

Several separation techniques are applicable for these functionalized dendrimers including precipitation, two-phase catalysis and immobilization of the dendrimer to insoluble support (polystyrene, silica). For every separation strategy the function of dendritic scaffold should be critically evaluated to really understand and appreciate the added value of the dendritic approach. This is obvious if the large size and the globular structure of dendrimer is utilized to facilitate separation by nanofiltration techniques or dialysis. Nanofiltration can be performed batchwise and in a continuous-flow membrane reactor (CFMR). The common problems involved in catalyst recycling also pertain to dendritic catalysts. These include dendrimer or catalyst decomposition, dendrimer leaching, metal leaching and catalyst deactivation. In addition, the binding strength of the catalyst to the support is also important; a low association constant will result in leaching of the noncovalently anchored catalyst. So far only a relatively few examples have been reported, and therefore it is too early to draw general conclusions. Further optimization of noncovalently supported catalysts and deeper understanding of the nature, the stability and the localization of the binding interactions between the catalyst and the support is required before application of these systems becomes realistic. However, it is clear from the examples reported so far that the supramolecular approach has advantages compared to covalent approaches. The preparation is simplified, and the dendrimer can be used as multi-purpose scaffold for a variety of catalytic applications. Also, as the binding of catalysts is reversible, the dendrimer can be defunctionalized enabling recycling of the dendritic scaffold. Many other new possibilities come to mind, but these still have to be brought into practice.

As already demonstrated for dendrimers with covalently attached catalysts, and also for noncovalently anchored systems, it seems that less well defined macromolecules such as hyperbranched polymers can be used equally well. They have the advantage that the preparation is more convenient and therefore the support is far cheaper. Well-defined macromolecular supports, such as "perfect" dendrimers, can be very helpful for mechanistic studies, validating future research in this area. In addition, in some reactions the perfect structure might lead to enhanced catalyst performance, in which case the use of dendritic catalysts could compete with their cheaper hyperbranched analogues. In general, the area of supramolecular dendritic catalysis is very promising as there are many clear advantages, but more studies are required to obtain the detailed knowledge facilitating the step towards commercial applications.

References

1. Astruc D, Chardac F (2001) Chem Rev 101:2991
2. Kreiter R, Kleij AW, Klein Gebbink RJM, van Koten G (2001) In: Vögtle F, Schalley CA (eds) Dendrimers IV: Metal Coordination, Self Assembly, Catalysis 217:163. Springer, Berlin Heidelberg New York
3. Oosterom GE, Reek JNH, Kamer PCJ, van Leeuwen PWNM (2001) Angew Chem Int Ed 40:1828
4. Crooks RM, Zhao M, Sun L, Chechik V, Yueng LK (2001) Acc Chem Res 34:181
5. Reek JNH, de Groot D, Oosterom GE, Kamer PCJ, van Leeuwen PWNM (2002) Rev Mol Biotechnol 90:159
6. Reek JNH, de Groot D, Oosterom GE, Kamer PCJ, van Leeuwen PWNM (2003) Comptes Rendus Chimie 6:106
7. van Heerbeek R, Kamer PCJ, van Leeuwen PWNM, Reek JNH (2002) Chem Rev 102:3717
8. Caminade A-M, Maraval V, Regis L, Majoral J-P (2002) Curr Org Chem 6:739
9. Newkome GR, Moorefield CN, Vögtle F (1996) Dendritic Molecules. Verlag Chemie, Weinheim
10. Vögtle F (1998) Top Curr Chem 197:1
11. Majoral J-P, Caminade A-M (1999) Chem Rev 99:845
12. Bosman AW, Janssen HM, Meijer EW (1999) Chem Rev 99:1665
13. Knapen JWJ, van der Made AW, de Wilde JC, van Leeuwen PWNM, Wijkens P, Grove DM, van Koten G (1994) Nature 372:659
14. Adams RD, Cotton FA (eds) (1998) Catalysis by Di- and Polynuclear Metal Cluster Complexes. Wiley, New York
15. Broussard ME, Juma B, Train SG, Peng W-J, Laneman SA, Stanley GG (1993) Science 260:1784
16. Ulman M, Grubbs RH (1999) J Org Chem 64:7202
17. Mukhopadhyah S, Rothenberg G, Gitis D, Sasson Y (2000) J Org Chem 65:3107
18. Van de Kuil LA, Grove DM, Gossage RA, Zwikker JW, Jenneskens LW, Drenth W, van Koten G (1997) Organometallics 16:4985
19. Sirkar KK, Shanbhag PV, Kovvali AS (1999) Ind Eng Chem Res 38:3715
20. Kragl U, Vasic-Racki D, Wandrey C (1992) Chem-Ing-Tech 64:499
21. Kragl U, Vasic-Racki D, Wandrey C (1993) Indian J Chem 32B:103
22. Bommarius AS (1993) In: Rehm H-J, Reed G, Pühler A, Stadler P, Stephanopoulos G (eds) Biotechnology, Bioprocessing, vol 3. VCH, Weinheim, p 427
23. Prazeres DMF, Cabral JMS (1994) Enzyme Mycrob Technol 16:738
24. Kragl U (1996) In: Godfrey T, West S (eds) Industrial Enzymology. MacMillan, Hampshire
25. Zeng F, Zimmerman SC (1997) Chem Rev 97:1681
26. De Groot D, de Waal BFM, Reek JNH, Schenning APHJ, Kamer PCJ, Meijer EW, van Leeuwen PWNM (2001) J Am Chem Soc 123:8453
27. van Heerbeek R (2006) Thesis, University of Amsterdam
28. Oosterom GE, Steffens S, Reek JNH, Kamer PCJ, van Leeuwen PWNM (2002) Top Catal 19:61
29. Chen R, Bronger RPJ, Kamer PCJ, van Leeuwen PWNM, Reek JNH (2004) J Am Chem Soc 126:14557
30. van der Veen LA, Keeven PH, Schoemaker GC, Reek JNH, Kamer PCJ, van Leeuwen PWNM, Lutz M, Spek AL (2000) Organometallics 19:872

31. Sandee AJ, Reek JNH, Kamer PCJ, van Leeuwen PWNM (2001) J Am Chem Soc 123:8468
32. van de Coevering R, Kuil M, Klein Gebbink RJM, van Koten G (2002) Chem Commun, p 1636
33. Ooe M, Murata M, Takahama A, Mizugaki T, Ebitani K, Kaneda K (2003) Chem Lett 32:692
34. Ooe M, Murata M, Mizugaki T, Ebitani K, Kaneda K (2004) J Am Chem Soc 126:1604
35. Zeng H, Newkome GR, Hill CL (2000) Angew Chem Int Ed 39:1771
36. Plault L, Hauseler A, Nlate S, Astruc D, Ruiz J, Gatard S, Neumann R (2004) Angew Chem Int Ed 43:2924
37. Sunder A, Heinemann J, Frey H (2000) Chem Eur J 6:2499
38. Schweb E, Mecking S (2001) Organometallics 20:5504
39. Schweb E, Mecking S (2005) Organometallics 24:3758
40. Slagt MQ, Stiriba S-E, Klein Gebbink RJM, Kautz H, Frey H, van Koten G (2002) Macromolecules 35:5734
41. Stiriba S-E, Slagt MQ, Kautz H, Klein Gebbink RJM, Thomann R, Frey H, van Koten G (2004) Chem Eur J 10:1267
42. Slagt MQ, Stiriba S-E, Kautz H, Klein Gebbink RJM, Frey H, van Koten G (2004) Organometallics 23:1525
43. Garcia-Bernabé A, Tzschucke CC, Bannwarth W, Haag R (2005) Adv Synth Catal 347:1389

Top Organomet Chem (2006) 20: 61–96
DOI 10.1007/3418_032
© Springer-Verlag Berlin Heidelberg 2006
Published online: 23 June 2006

Stereoselective Dendrimer Catalysis

Jutta K. Kassube · Lutz H. Gade (✉)

Anorganisch-Chemisches Institut, Universität Heidelberg, Im Neuenheimer Feld 270,
69120 Heidelberg, Germany
lutz.gade@uni-hd.de

Abstract Enantioselection in a stoichiometric or catalytic reaction is governed by small increments of free enthalpy of activation, and such transformations are thus in principle suited to assessing "dendrimer effects" which result from the immobilization of molecular catalysts. Chiral dendrimer catalysts, which possess a high level of structural regularity, molecular monodispersity and well-defined catalytic sites, have been generated either by attachment of achiral complexes to chiral dendrimer structures or by immobilization of chiral catalysts to non-chiral dendrimers. As monodispersed macromolecular supports they provide ideal model systems for less regularly structured but commercially more viable supports such as hyperbranched polymers, and have been successfully employed in continuous-flow membrane reactors. The combination of an efficient control over the environment of the active sites of multi-functional catalysts and their immobilization on an insoluble macromolecular support has resulted in the synthesis of catalytic dendronized polymers. In these, the catalysts are attached in a well-defined way to the dendritic sections, thus ensuring a well-defined microenvironment which is similar to that of the soluble molecular species or at least closely related to the dendrimer catalysts themselves.

Keywords Dendrimers · Stereoselective catalysis · Homogeneous catalysis ·
Heterogeneous catalysis

Abbreviations

ALB	Al-Li-bis(binaphthoxide)
BICOL	$9H,9'H$-[4,4′]Bicarbazole-3,3′-diol
BINAP	2,2′-Bis(diphenylphosphino)-1,1′-binaphthyl
BINOL	1,1′-Bi-2-naphthol
COD	1,5-Cyclooctadiene
Danishefsky's diene	1-Methoxy-3-(trimethylsilyloxyl)buta-1,3-diene
DPEN	1,2-Diphenylethylenediamine
ee	Enantiomeric excess
G_x	Generation x (1, 2, 3, …)
Josiphos	1-[2-(Diphenylphosphino)ferrocenyl]ethyldicyclohexyl-phosphine
NOBIN	2-Amino-2′-hydroxy-1,1′-binaphthyl
PAMAM	Poly(amido)amine
PPI	Poly(propyleneimine)
Pyrphos	3,4-Bis(diphenylphosphino)pyrrolidine
salen	Ethylenebis(salicylimine)
TADDOL	2,2-Dimethyl-$\alpha,\alpha,\alpha',\alpha'$-tetraphenyl-1,3-dioxolane-4,5-dimethanol

1
Introduction

Homogeneous chiral catalysts are well-defined molecular systems that may
display high activity and selectivity in catalytic reactions, combined with an
excellent reproducibility of the experimental results. The production of chiral enantiomerically pure catalysts is expensive and thus catalyst recycling is
an important practical objective. This challenge was recognized early on and
has led to the development of supported molecular catalysts and, more recently, dendrimer catalysts [1, 2]. With the latter, catalyst recycling has been
achieved using membrane reactors as well as catalyst precipitation and subsequent filtration, although frequently with deteriorating catalyst performance
over time. This utilitarian aspect of dendrimer catalysis has provided the
motivation for much of the work on chiral dendrimer catalysts. These have
been generated either by attachment of achiral complexes to chiral dendrimer
structures or by immobilization of chiral catalysts to non-chiral dendrimers,
by methods described in more detail elsewhere in this monograph.

Dendrimer fixation may be achieved by attachment of catalysts at the periphery of dendrimers (Fig. 1a) in the way first established by van Koten,
van Leeuwen and co-workers in their pioneering work on the Karasch reaction [3, 4]. The second possibility is the attachment of one or more dendritic
wedges to the catalysts, which are then located at the core of the result-

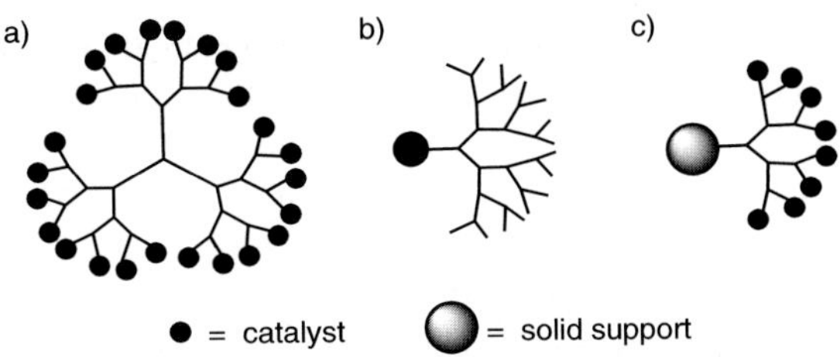

Fig. 1 Fixation of catalytic metal centres (represented by the *black spheres*) in exodendral (**a**) and endodendral (**b**) positions of dendrimers or on dendronized polymers (**c**)

ing functionalized dendrimers (Fig. 1b). The first example of such catalysts was reported by Brunner et al. ("dendrizymes"), who studied the influence of a chiral dendritic periphery on the performance of asymmetric cyclopropanation catalysts [5, 6]. Recently, a third type of chiral dendritic catalyst has been developed which is based on an insoluble polymer support loaded with dendritic wedges. These are functionalized by attachment of molecular catalysts to their periphery (Fig. 1c). The structures obtained following this strategy combine the well-defined dendritic architecture near the active sites of the catalytic phase with the ease of reuse characteristic of common heterogeneous catalysts.

Since enantioselection in a stoichiometric or catalytic reaction is governed by small increments of free enthalpy of activation [7], such transformations are particularly suited to assessing "dendrimer effects" which result from the immobilization of catalysts. In the assessment of such multi-site macromolecular catalysts, it is essential to establish whether the immobilized catalyst units retain their identity and are not altered by the nature of the dendrimer backbone. The linker and spacer units employed in the fixation of the catalysts may be crucial in this respect as well as the functional groups present in the dendrimer. Regarding the dendrimer core structure itself, the length and conformational rigidity of the branches and spacers are important factors when evaluating a dendritic catalyst. For immobilized asymmetric catalysts, even subtle conformational changes may significantly influence their stereoselectivity. The interplay of all these factors will generally determine detrimental or beneficial dendrimer effects on catalyst performance, which will be the focus of this overview of the field of asymmetric dendrimer catalysis.

2
Dendrimer Fixation of Chiral Catalysts

The underlying concept of Brunner's dendrizymes [5, 6] was the creation of a chiral dendritic architecture in the environment of a catalytic centre, which should induce stereoselectivity in the same way as in enzymes. In metalloenzymes, in particular, the first coordination sphere of the metal centre is frequently achiral (see for instance the ubiquitous tris-histidine binding sites!), and the chiral induction is due to the chirality of the polypeptide protein structure which allows for chiral secondary interactions with the substrates. In contrast to the biological systems, the reaction performance of dendrimers containing a chiral dendritic backbone has been found to be unsatisfactory [8, 9]. This is thought to be due to the fact that the dendrimers that have been employed do not form well-defined secondary structures in the same way as polypeptides. Consequently, the arrangement of functional groups in the "second sphere" of the catalytic site, and thus their interaction with a potential substrate, remains ill-defined.

In view of the inefficiency of the chiral induction effected by chiral dendritic structures, most attention has been directed towards the covalent or electrostatic [10] fixation of established chiral mononuclear complexes to achiral dendritic cores or wedges. A number of efficient chiral dendrimer catalysts have been obtained, as will be discussed in the following sections. However, even in these cases particular care has to be taken to avoid the negative interference of functional groups in the dendrimer core with the attached catalytic sites. This aspect is thought to be important in the enantioselective ethylation of benzaldehyde with chiral dendrimer catalysts, for which both the chemical yields and the enantiomeric excesses were found to decrease with increasing generation of the dendrimers [8, 9]. Multiple interactions of the catalytic sites on the dendritic surface with the substrate at higher generations add to the observed negative effects. To relieve these interactions with the dendrimer end groups at the periphery of higher generations, the introduction of an alkyl chain as a spacer or the use of a rigid hydrocarbon backbone have been proposed [11].

3
Chiral Catalytic Sites at the Periphery of Dendrimers

3.1
Asymmetric Catalysis with Immobilized Phosphine-Based Catalysts

The first example of asymmetric rhodium-catalyzed hydrogenation of prochiral olefins in dendrimer catalysis was reported by Togni et al., who immobilized the chiral ferrocenyl diphosphine "Josiphos" at the end groups of dendrimers, thus obtaining systems of up to 24 chiral metal centres in the periphery (Fig. 2) [12–14]. The fact that the catalytic properties of the dendrimer catalysts were almost identical to those of the mononuclear catalysts was interpreted as a manifestation of the independence of the individual catalytic sites in the macromolecular systems.

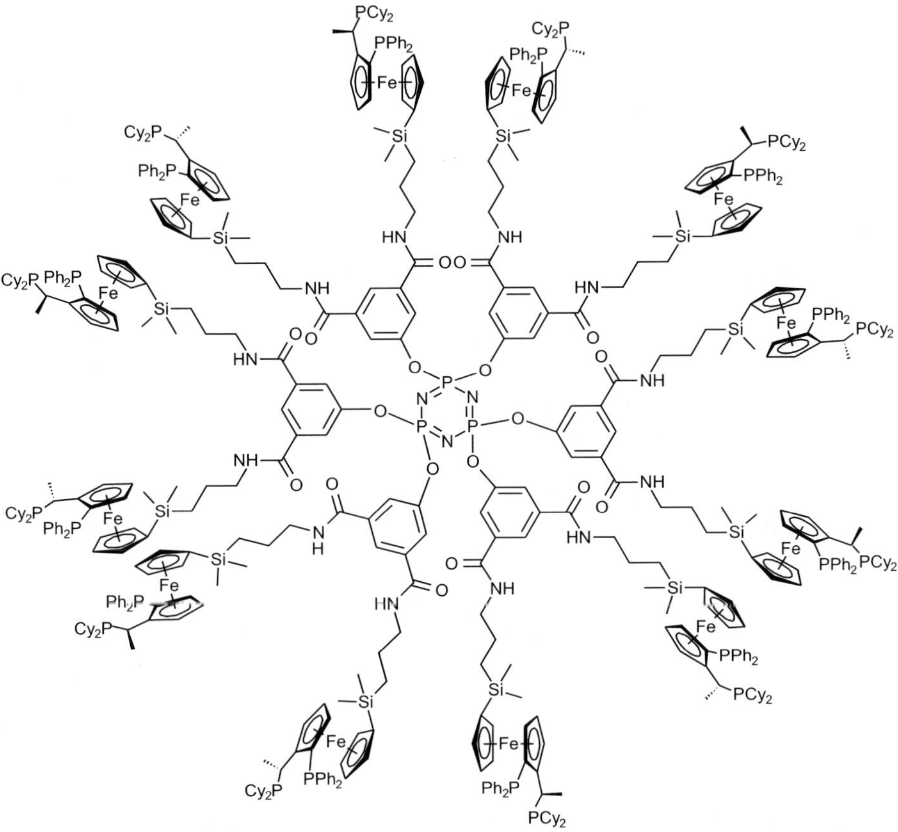

Fig. 2 Fixation of chiral "Josiphos" ligands to a first-generation dendrimer containing a cyclotriphosphazene core

In a comprehensive study carried out by the same group, the asymmetric hydrogenation of dimethyl itaconate, asymmetric allylic substitutions and asymmetric hydroboration reactions catalyzed by the multi-Josiphos rhodo-dendrimers were investigated [15]. The stereoselectivities obtained with the monodisperse dendrimer catalysts with up to 16 metal centres in the periphery were found to be very similar to those of the mononuclear reference systems. The authors concluded that the absence of negative dendrimer effects is probably the best possible result when cooperativity effects between single catalyst units do not play any relevant role. Small losses in selectivity, observed upon going to higher dendrimer generations, may be due to local concentration effects which become important, in particular, for cationic catalyst species.

A series of chiral phosphine-functionalized poly(propyleneimine) (PPI) and poly(amido)amine (PAMAM) dendrimers was synthesized by reaction of carboxyl-linked C_2-chiral Pyrphos ligands (Pyrphos = 3,4-bis(diphenyl-phosphino)pyrrolidine) with 0th–5th-generation PPI (0th–4th-generation PAMAM) using ethyl-N,N-dimethylaminopropyl carbodiimide (EDC)/1-hydroxybenzotriazole as a coupling reagent (Scheme 1) [16]. The functionalized dendrimers were characterized by NMR spectroscopy, elemental analysis, and FAB and MALDI-TOF mass spectrometry, thus establishing their molecular masses of up to 20 700 amu for the PPI derivatives.

Metallation of the multi-site phosphines with $[Rh(COD)_2]BF_4$ (COD = 1,5-cyclooctadiene) or $[PdCl_2(NCCH_3)_2]$ cleanly yielded the cationic rhodo-dendrimers or palladodendrimers, respectively, containing up to 64 metal centres (Scheme 2).

The relationship between the size/generation of the rhododendrimers and their catalytic properties was established inter alia in the asymmetric hydrogenation of Z-methyl-α-acetamidocinnamate and dimethyl itaconate. Generally, a decrease in activity of the dendrimer catalysts was observed on going to the higher generations for the PPI- and PAMAM-based dendrimers (Fig. 3).

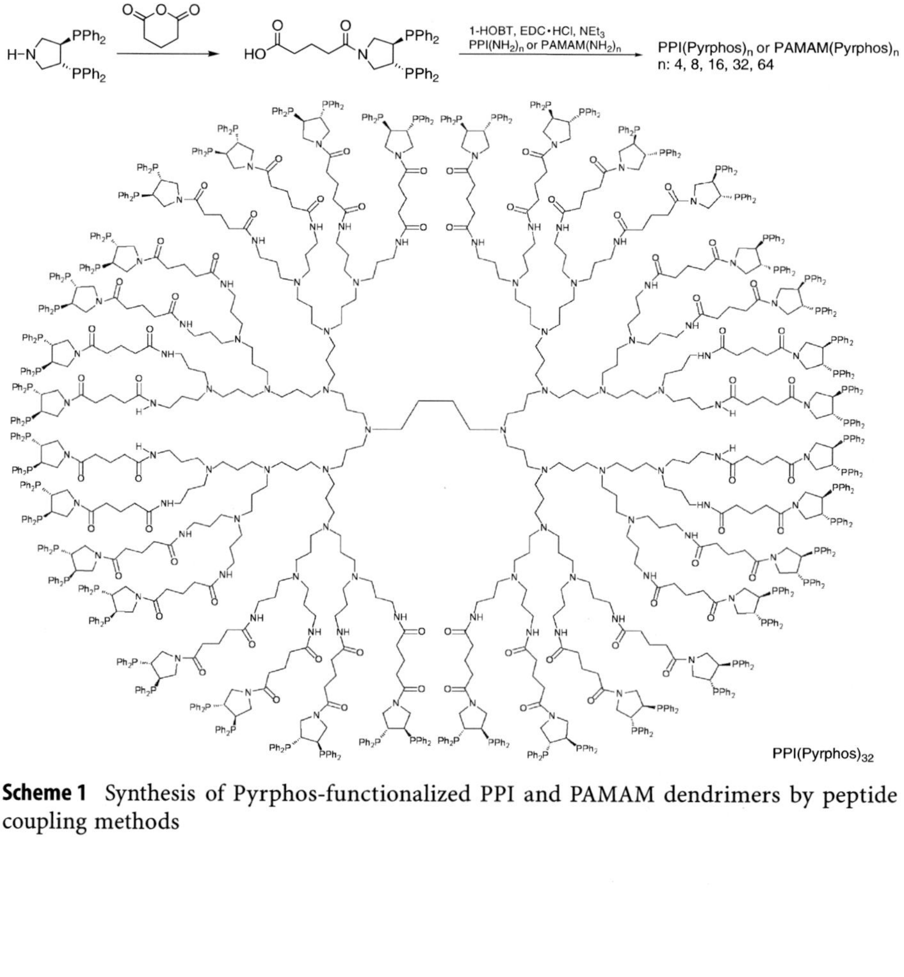

Scheme 1 Synthesis of Pyrphos-functionalized PPI and PAMAM dendrimers by peptide coupling methods

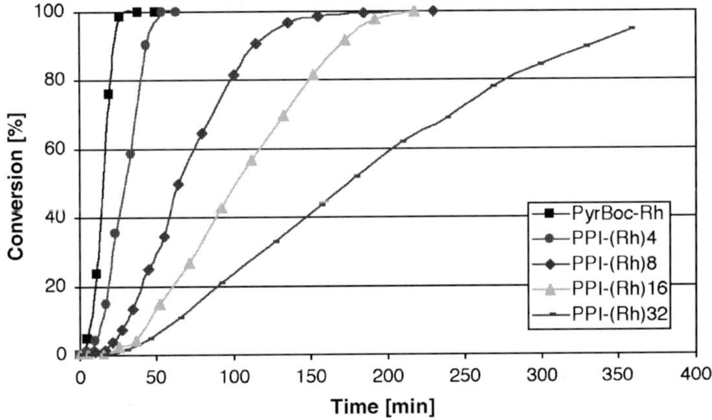

Scheme 2 Metallation of the Pyrphos-functionalized PPI and PAMAM dendrimers

Fig. 3 Conversion curves for the asymmetric hydrogenation of Z-methyl-α-acetamido-cinnamate for the different catalyst generations

However, for the Rh(COD) systems, the enantioselectivity in the hydrogenation of Z-methyl-α-acetamidocinnamate remained barely affected (Fig. 4) or decreased only slightly, depending on the chosen substrate.

In an effort to extend the use of the Pyrphos-derived dendrimers to asymmetric Pd-catalyzed coupling reactions, strongly positive selectivity effects were observed upon going to very large multi-site chiral dendrimer catalysts. This enhancement of the catalyst selectivity was observed in palladium-catalyzed allylic substitutions, such as that displayed in Scheme 3, which are known to be particularly sensitive to small changes in the chemical environment of the active catalyst sites [17].

The mononuclear catalyst [(Boc-Pyrphos)PdCl$_2$], which is very unselective for this transformation [9% enantiomeric excess (ee)], provided the point of reference for the subsequent studies with the dendrimer catalysts. This system and the metalladendrimers PPI(PyrphosPdCl$_2$)$_4$ – PPI(PyrphosPdCl$_2$)$_{64}$

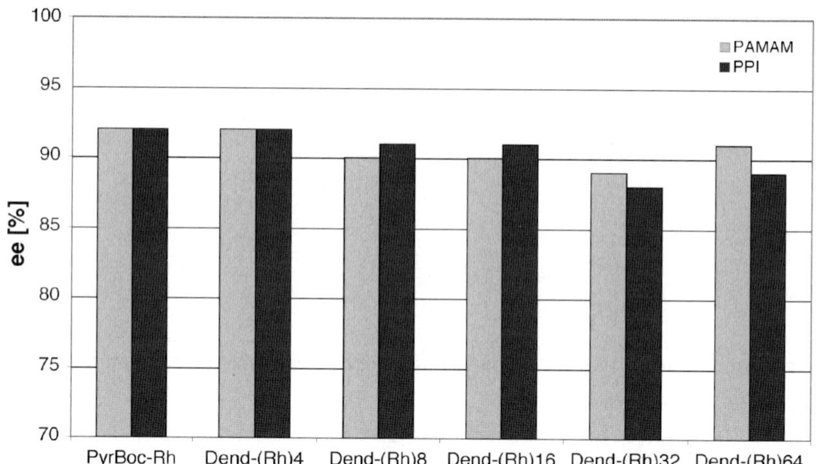

Fig. 4 Enantioselectivities for the different PPI- and PAMAM-derived catalyst generations in the hydrogenation of Z-methyl-α-acetamidocinnamate

Scheme 3 Asymmetric allylic amination of 1,3-diphenyl-1-propene-3-acetate catalyzed by Pyrphos-palladium complexes

and $PAMAM(PyrphosPdCl_2)_4$ – $PAMAM(PyrphosPdCl_2)_{64}$ in 0.3 mol % catalyst concentration were studied in the catalytic amination of 1,3-diphenyl-1-propene-3-acetate and gave the results displayed in Fig. 5.

A remarkable increase in catalyst selectivity was observed as a function of the dendrimer generation. This steady increase in ee values for the allylic amination was less pronounced for the PPI-derived catalysts [40% ee for $PPI(PyrphosPdCl_2)_{64}$] than for the palladium-PAMAM dendrimer catalysts, for which an increase in selectivity from 9% ee for the mononuclear complex to 69% ee for $PAMAM(PyrphosPdCl_2)_{64}$ was found. The same general trend was observed in the asymmetric allylic alkylation of 1,3-diphenyl-1-

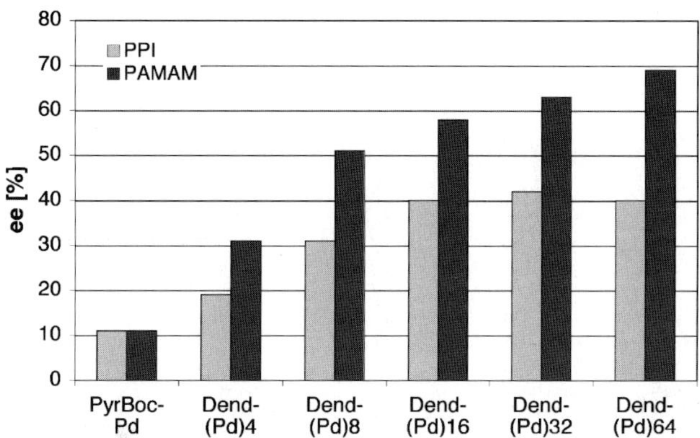

Fig. 5 Dependence of the enantiomeric excesses found for the allylic amination in Scheme 3 on the dendrimer generation for both precatalyst series PPI(PyrphosPdCl$_2$)$_4$ – PPI(PyrphosPdCl$_2$)$_{64}$ and PAMAM(PyrphosPdCl$_2$)$_4$ – PAMAM(PyrphosPdCl$_2$)$_{64}$

propene-3-acetate with sodium dimethylmalonate, which indicates that the results of the amination reaction may be typical for allylic substitutions in general [18–22].

The underlying mechanistic reasons for this strongly positive "dendrimer effect" are thought to be based on a dismutation reaction of the Pyrphos-palladium complexes, giving {(Pyrphos)$_2$Pd} sites which act as catalytic centres upon going to higher dendrimer generations. The latter effect is thought to be due to the high local catalyst concentration enforced by the attachment of the complexes to the dendrimer supports. The same selectivity effects were observed for the mononuclear species upon addition of a second equivalent of Pyrphos ligand.

3.2
Asymmetric Catalysis with Immobilized Non-Phosphine-Based Catalysts

The majority of studies into the catalytic behaviour of dendrimers with chiral catalytic centres at the periphery of the dendritic support have concerned non-phosphine-based catalysts. As has become apparent in these studies, the effect of the dendrimer fixation on the catalytic performance generally depends on the individual system. Factors such as the high local density of catalytic sites, the interaction of functional groups in the dendrimer backbone with the catalysts and the structural rigidity or flexibility of the dendrimers seem to play a role in many cases.

Jacobsen et al. reported a spectacular example of dendrimer-induced rate enhancement. They synthesized dendrimer-bound [Co^{II}(salen)] complexes (salen = ethylenebis(salicylimine)) with up to 16 catalytic sites, using PAMAM dendrimers as supporting materials (Fig. 6) [23]. In the hydrolytic kinetic

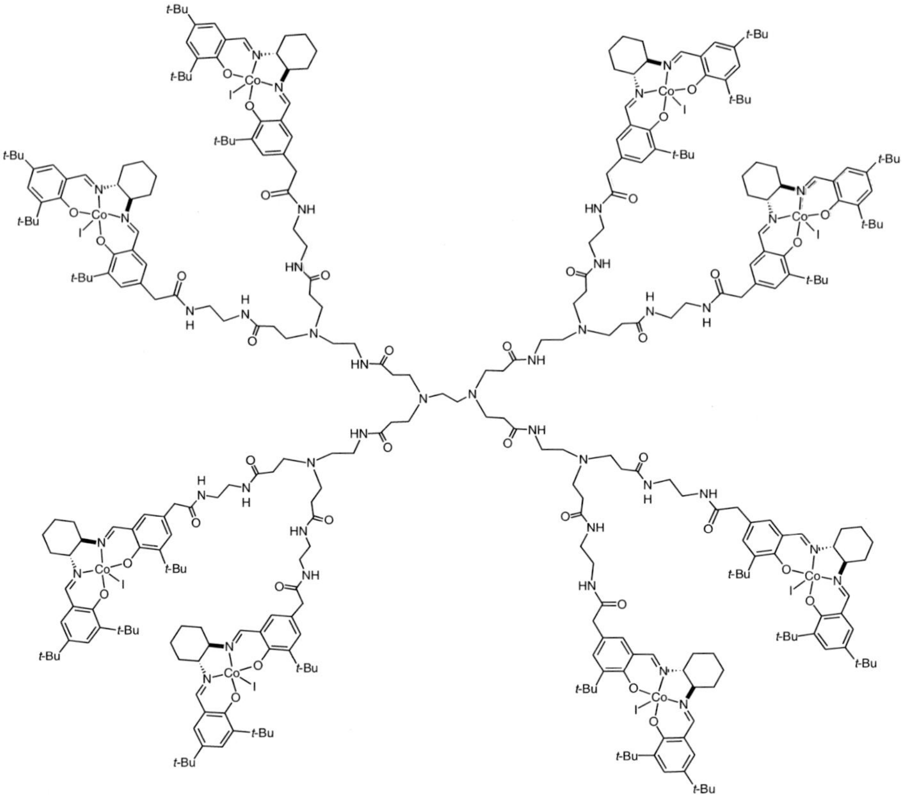

Fig. 6 The dendrimer-bound [Co(salen)] complexes studied by Breinbauer and Jacobsen

resolution (HKR) of terminal epoxides these catalysts exhibit significantly enhanced catalytic activity, in comparison with that of the mononuclear system. As had been shown previously, the kinetics of this reaction are second order in the concentration of the [Co(salen)] complex, a fact which was explained by a cooperative two-site mechanism. The dendrimer effect may therefore be attributed to the restricted conformation imposed by the dendrimer structure and the increase in the local effective molarity of [Co(salen)] units.

Soai et al. developed poly(amidoamine), poly(phenylethyne) and carbosilane dendrimers containing chiral β-amino alcohols on their periphery (Fig. 7) [11, 24–28]. These systems were studied as catalysts for the enantioselective addition of dialkylzinc reagents to aldehydes and *N*-diphenylphosphinylimines. The different kinds of supporting dendritic structures gave rise to varying catalytic activities and selectivities. The functionalized PAMAM dendrimers of generations 0 and 1, bearing four and eight sites of chiral amino alcohols, catalyzed the addition of dialkylzinc derivatives to *N*-diphenylphosphinylimines with only moderate enantioselectivity com-

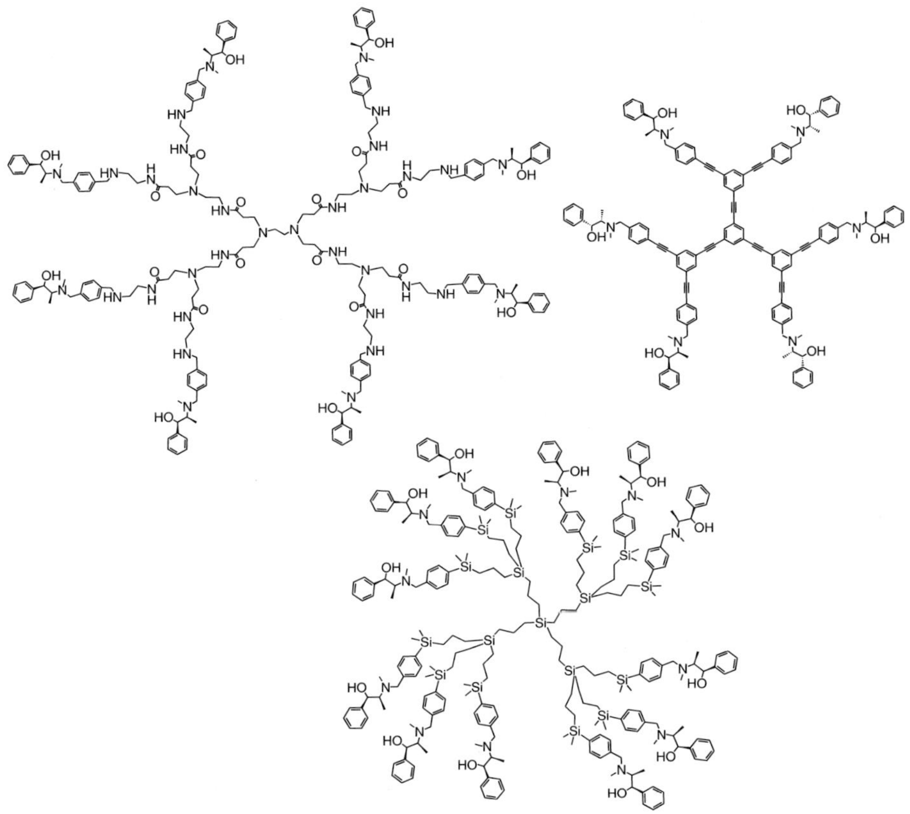

Fig. 7 The dendrimer-immobilized chiral amino alcohols studied by Soai et al. in the asymmetric alkylation of prochiral carbonyl compounds with dialkylzinc reagents

pared to the monomeric system. To obtain appreciable catalytic conversions, an excess of the zinc reagent was needed in each case, which is due to the fact that the nitrogen and oxygen atoms of the PAMAM skeleton coordinate to the zinc. The authors concluded that this led to a change of the conformation of the chiral dendrimers, which was thought to be the reason for the subsequent decrease in selectivity.

In order to avoid this unfavourable effect of the functional groups in the dendrimer structure, a rigid hydrocarbon backbone without heteroatoms was synthesized. Dendrimers with poly(phenylethyne) backbones, bearing three and six ephedrine derivatives at the periphery, were studied in the alkylation of aldehydes and N-diphenylphosphinylimines and proved to be highly enantioselective catalysts. For example, the system containing six catalytic sites catalyzed the addition of diisopropylzinc to aldehydes with enantioselectivities of up to 86% ee. As a third backbone a polycarbosilane dendrimer was used, which is chemically inert and more flexible than the poly(phenylethyne) structure. The chiral dendrimers, bearing four and 12 ephedrine moieties catalyzed the reaction of diisopropylzinc with 3-phenylpropanal with even higher enantioselectivities (up to 93% ee).

Chan et al. synthesized first- and second-generation dendrimers containing up to 12 chiral diamines at the periphery (Fig. 8) [29]. Their ruthenium(II) complexes displayed high catalytic activity and enantioselectivity in the asymmetric transfer hydrogenation of ketones and imines. Quantitative yields, and in some cases a slightly higher enantioselectivity compared to those of the monomeric systems (up to 98.7% ee), were obtained.

In 2002, Sasai et al. reported the synthesis of dendritic heterobimetallic multi-functional chiral catalysts, containing up to 12 1,1′-bi-2-naphthol (BINOL) units at their terminal positions (Fig. 9) [30]. On treating these functionalized dendrimers with AlMe$_3$ and n-BuLi, insoluble metallated Al-Li-bis(binaphthoxide) generation x (G$_x$-ALB) catalysts were obtained, which showed moderate catalytic activity in the asymmetric Michael reaction of 2-cyclohexenone with dibenzyl malonate (Scheme 4).

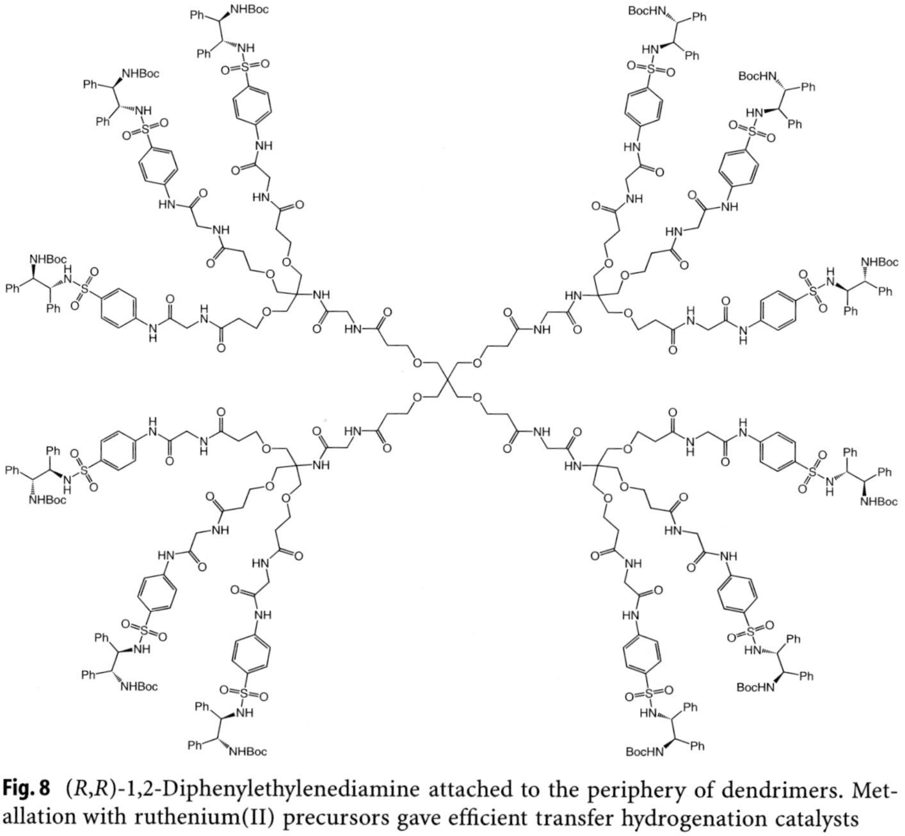

Fig. 8 (*R,R*)-1,2-Diphenylethylenediamine attached to the periphery of dendrimers. Metallation with ruthenium(II) precursors gave efficient transfer hydrogenation catalysts

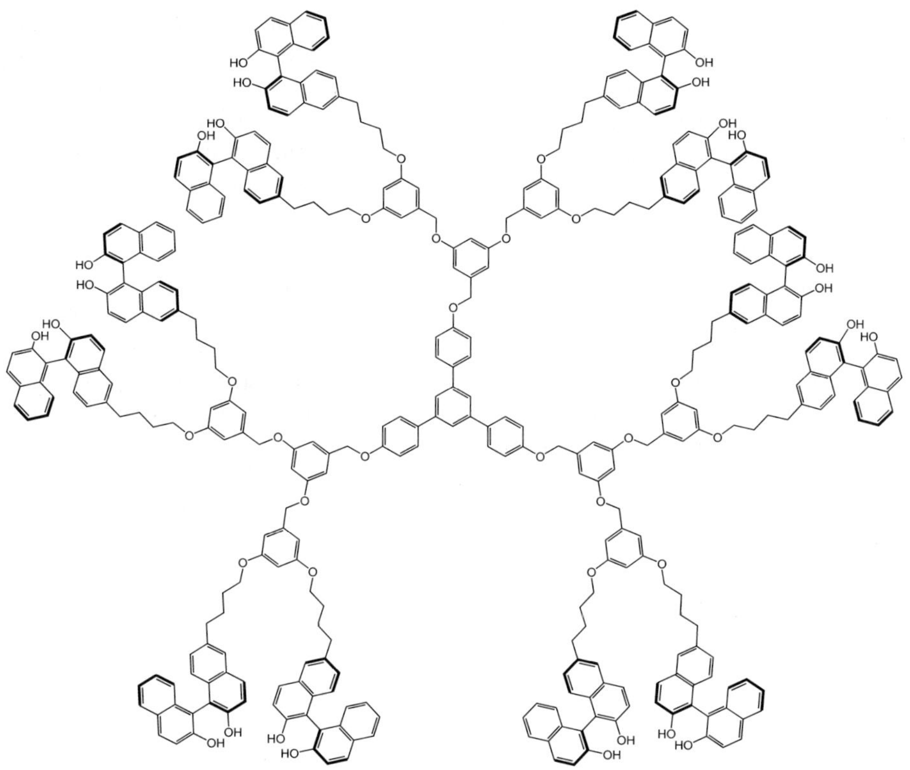

Fig. 9 Chiral BINOL ligands attached to the periphery of a poly(aryl ether) dendrimer

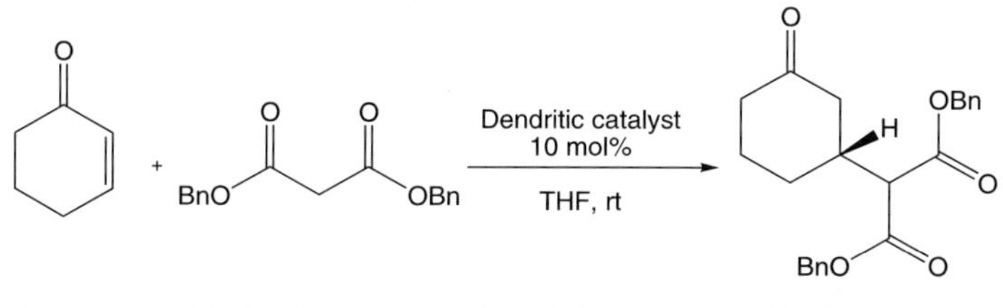

Scheme 4 Asymmetric Michael reaction of 2-cyclohexenone with dibenzyl malonate catalyzed by the metallated dendrimer shown in Fig. 9

Using the first-generation dendritic ALB as catalyst, the Michael adduct was obtained with 91% ee and in 63% yield after 48 h. Under similar conditions, the G_2 dendritic ALB gave the product with 91% ee in 59% yield. The dendritic catalysts could be recycled and reused twice, giving comparable results. It is notable that a catalyst derived from randomly introduced BINOLs on polystyrene resin only gave an essentially racemic product.

Another multi-centred BINOL derivative used in asymmetric Lewis acid catalysis should be mentioned in this context. In 2002, Chow et al. reported the preparation of the G_0 and G_1 generations of chiral 1,1'-binaphthalene-based dendritic ligands, using an oligo(arylene) framework as rigid supporting material (Fig. 10) [31]. Their corresponding aluminium complexes were shown to induce slightly higher reactivity and enantioselectivity than those of a monomeric 1,1'-binaphthalene catalyst in the Diels–Alder reaction between cyclopentadiene and 3-[(E)-but-2-enoyl]-oxazolidin-2-one. In the absence of intramolecular interactions among the catalytic centres, the catalyst reactivity and reaction enantioselectivity were found to be independent of the dendrimer generation.

Recently, Majoral et al. described the synthesis of a third-generation phosphorus-containing dendrimer possessing 24 chiral iminophosphine end groups derived from (2S)-2-amino-1-(diphenylphosphinyl)-3-methylbutane (Fig. 11) [32]. The dendritic catalyst was tested in allylic substitution reactions, using *rac*-(E)-diphenyl-2-propenyl acetate or pivalate as substrates. The observed enantioselectivities were good to excellent (max. 95% ee) in all reactions. After completion of the catalytic reaction, the catalyst could be reused at least twice after precipitation and filtration. A slight decrease

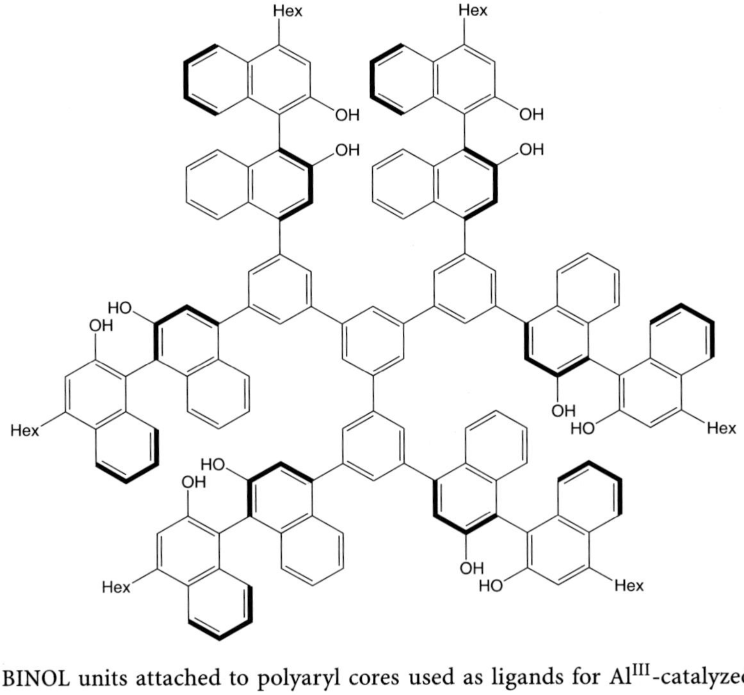

Fig. 10 BINOL units attached to polyaryl cores used as ligands for Al^{III}-catalyzed asymmetric Diels–Alder reactions

of the enantioselectivity was observed (first reuse 94% ee, second reuse 92% ee) and, as shown in Fig. 12, a diminished activity was found for the third run.

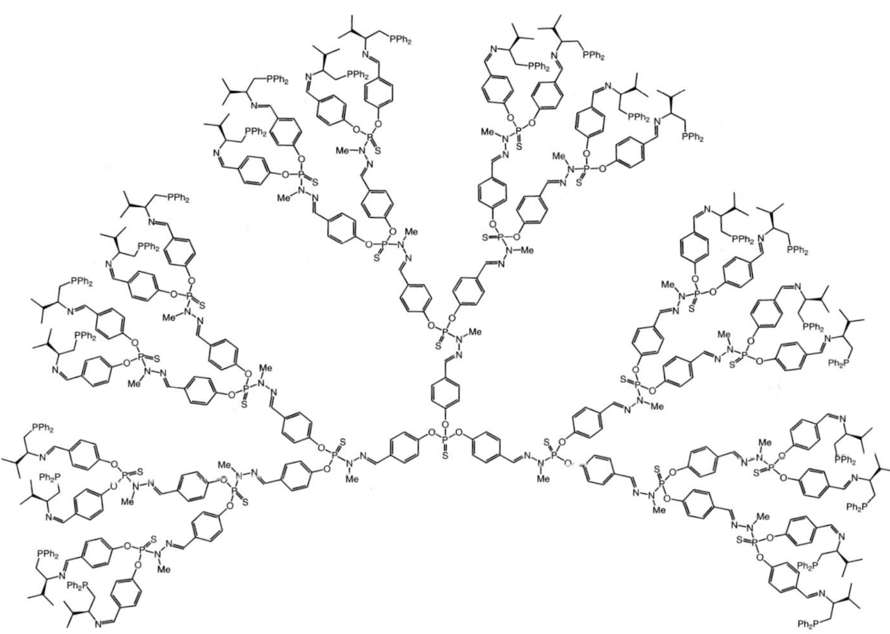

Fig. 11 The chiral dendrimers used by Majoral et al. for enantioselective allylic substitution catalyzed by palladium

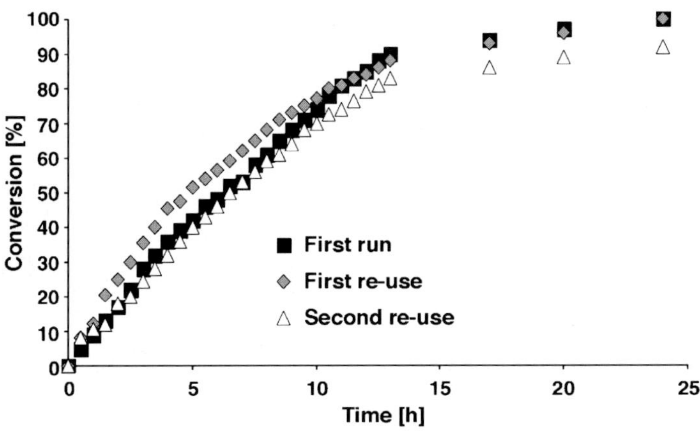

Fig. 12 Asymmetric allylic substitution catalyzed by the palladated dendrimer shown in Fig. 11 [32]

3.3
A Special Case: Enantioselective Borohydride Reduction of Ketones in the Periphery of Chiral Glycodendrimers

In all the examples of exodendrally functionalized enantioselective dendrimer catalysts, the active sites in the periphery of the support were well-defined immobilized molecular catalysts. An alternative is provided by the possibility of attaching chiral multi-functional molecules to the end groups of dendrimers which, due to their high local concentrations, may interact more or less strongly with an achiral reagent and thus induce enantioselectivity in a transformation of a prochiral substrate. Asymmetric induction thus occurs by way of a chiral functionalized microenvironment for a given reaction.

An interesting example of this kind of stereoselective dendrimer catalysis has been reported by Rico-Lattes et al., who prepared glycoden-

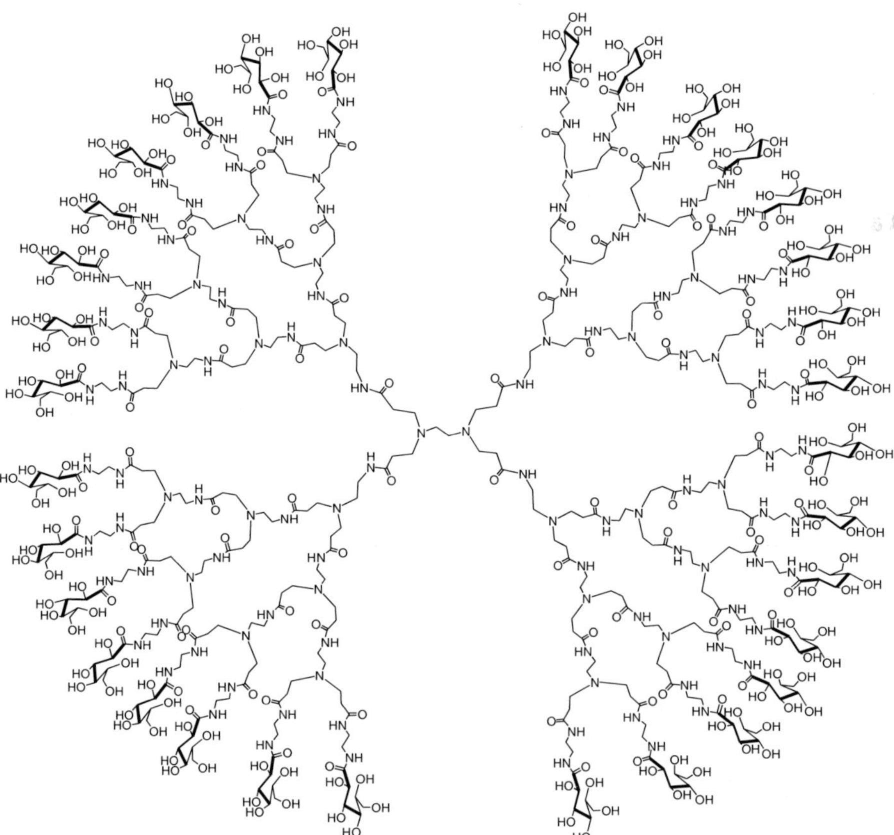

Fig. 13 Third-generation glycol-PAMAM dendrimers studied as microenvironments for the enantioselective reduction of prochiral ketones with sodium borohydride

drimers based on poly(amido)amines of generations 1–4 and gluconolactone (Fig. 13) [33–37]. These glycodendrimers were examined as catalysts for the reduction of prochiral ketones, using sodium borohydride as reducing agent. The corresponding chiral alcohols were isolated in high yields and had enantiopurities of up to 100% ee.

In water, the highest stereoselectivities were obtained by using the fourth-generation amphiphilic dendrimer, whereas, under heterogeneous reaction conditions in THF as solvent, the third-generation dendritic catalyst proved to be the most selective.

To explain this different behaviour, mechanistic studies of the system were performed with the aid of molecular modelling, ^{13}C NMR spectroscopy, induced circular dichroism, a systematic variation of the reaction parameters, and variation of the molecular structure of the sugar moieties and of the linking units. These studies established that under homogeneous reaction conditions (water), the main factor influencing the enantioselectivity is probably the ordering and specific orientation of the ketone at the chiral interface. Under heterogeneous conditions in THF the situation appears to be more complex.

4
Asymmetric Catalysis in the Core of Dendrimers:
Catalysts Attached to Dendritic Wedges

Brunner's concept of attaching dendritic wedges to a catalytically active metal complex represented the first example of asymmetric catalysis with metal complex fragments located at the core of a dendritic structure [5, 6]. Important early examples of catalysts in core positions were Seebach's TAD-DOL systems (TADDOL = 2,2-dimethyl-α,α,α',α'-tetraphenyl-1,3-dioxolane-4,5-dimethanol) [38, 39]. In general, the catalytic performance of such systems was either unchanged with respect to the simple mononuclear reference system or significantly lower. In no case has the potential analogy of this core fixation and the existence of efficient reactive pockets in enzymes been vindicated. This may be due to the absence of defined secondary structures in the dendrimers that have been employed to date.

4.1
Chiral BINAP- and BINOL-Based Dendrimer Catalysts

Two interesting reports by Chan and co-workers of dendritic core-functionalized Ru-BINAP (BINAP = 2,2′-bis(diphenylphosphino)-1,1′-binaphthyl) catalysts, which were employed in asymmetric hydrogenations and which were fully recoverable, have appeared recently [40, 41]. In particular, such dendrimers containing long alkyl chains in the periphery were synthesized and employed for asymmetric hydrogenation using a mixture of ethanol/hexane as reaction medium (Fig. 14).

This binary solvent system provided complete miscibility of the phases over a broad range of reaction temperatures and avoided the use of water during the catalytic conversion, with its established negative effects on the enantioselectivity [41]. Phase separation after complete reaction was induced by the addition of small quantities of water, and the recycling of the catalyst could be readily achieved. Remarkably, attachment of the dendrimer to the BINAP system did not lead to a decrease in selectivity.

In a related approach, Fan et al. synthesized a series of dendritic BINAP-Ru/chiral diamine ((R,R)-1,2-diphenylethylenediamine; DPEN) catalysts for the asymmetric hydrogenation of various simple aryl ketones (Fig. 15) [42]. The resulting systems displayed high catalytic activity and enantioselectivity and allowed facile catalyst recycling. In the case of 1-acetonaphthone and

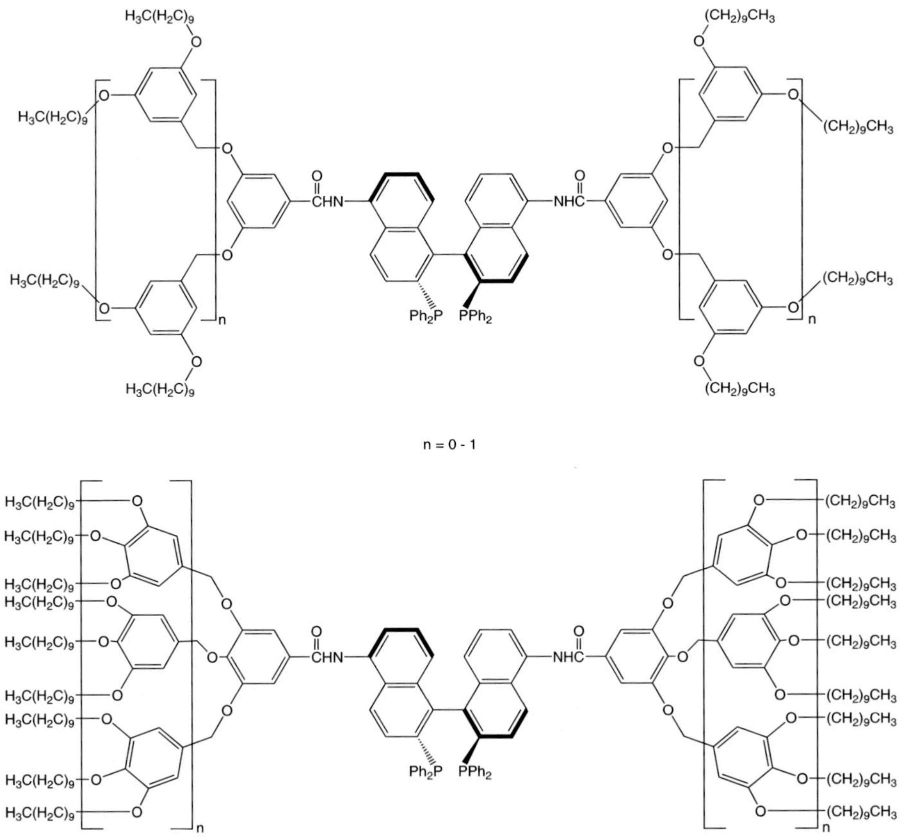

Fig. 14 BINAP core-functionalized dendrimers containing long alkyl chains in the periphery

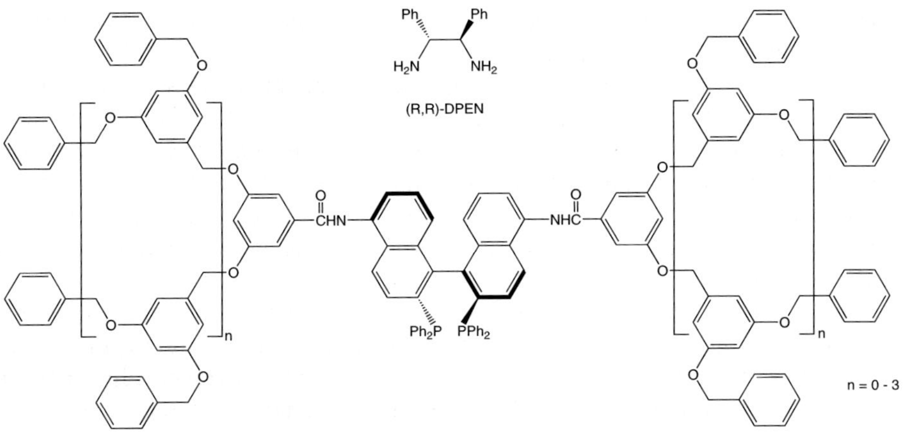

Fig. 15 BINAP ligand functionalized with poly(aryl ether) dendritic wedges

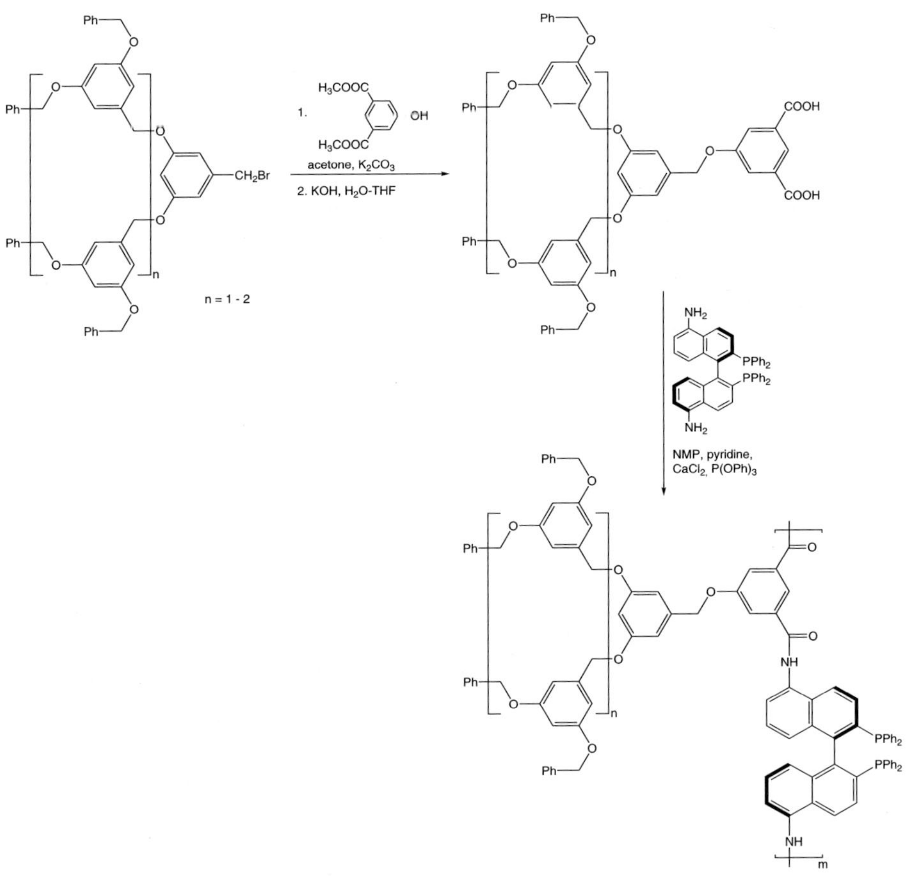

Scheme 5 The dendronized polymeric BINAP ligands studied by Fan et al.

$2'$-methylacetophenone, ee values of up to 95% were observed, which are comparable to the enantioselectivity reported by Noyori under similar conditions and higher than those of the heterogeneous poly(BINAP)-Ru catalyst reported by Pu and co-workers [43].

The same group also developed optically active dendronized polymeric BINAP ligands (see also Sect. 5) as a new type of macromolecular chiral catalyst for asymmetric hydrogenation. They could be synthesized by condensation of $5,5'$-diamino-BINAP with dendritic dicarboxylic acid monomers (Scheme 5) [44].

These polymeric Ru(BINAP) catalysts exhibited high catalytic activity and enantioselectivity (up to 92%) in the hydrogenation of simple aryl ketones, which is very similar to the results obtained with the corresponding parent Ru(BINAP) as well as the Ru(BINAP)-cored dendrimers referred to above. Unsurprisingly, they found that the pendant dendritic wedges have a major impact on the solubility and the catalytic properties of the polymeric catalysts, which could be easily recovered from the reaction mixture by simple precipitation.

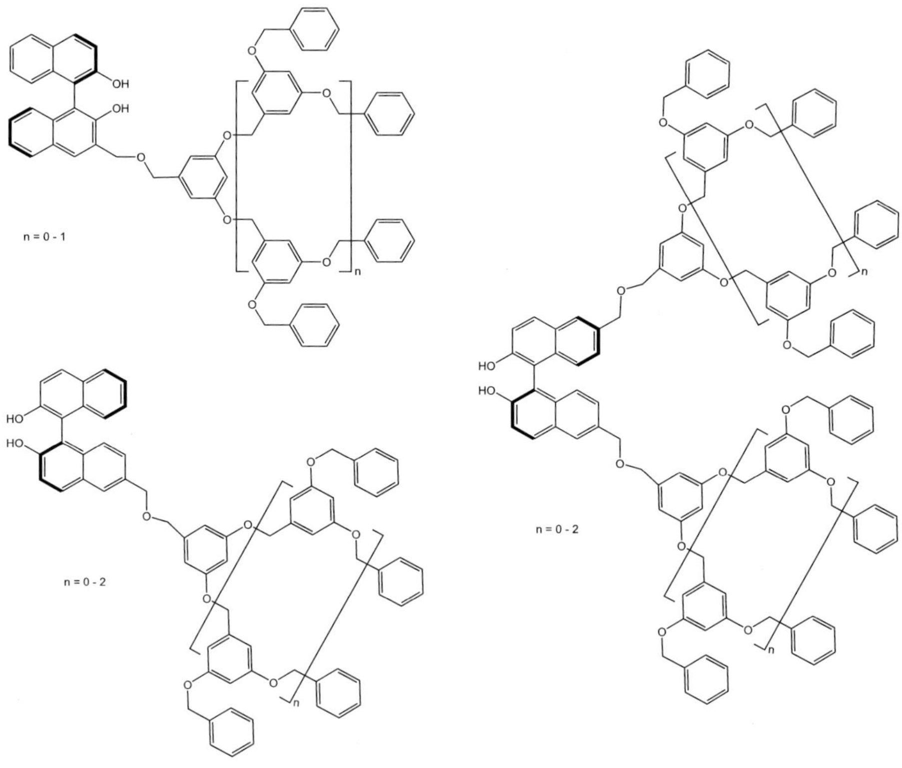

Fig. 16 BINOL ligands bearing dendritic poly(aryl ether) wedges

In 2003 Fan et al. synthesized three types of new chiral BINOL ligands (Fig. 16), bearing dendritic wedges located at the 3-, 6,6'- and 6-positions of the binaphthyl backbone in order to study the effect of the linking position and generation of the dendritic wedges on the catalyst properties [45, 46]. These new ligands were tested in the enantioselective Lewis acid catalyzed addition of diethylzinc to benzaldehyde, and high conversions (up to 99%) and enantioselectivities were observed. A marked effect of the positions of attachment for the linkers as well as the dendron generation on the enantioselectivity and/or activity was found for all three types of dendritic catalysts. Among these systems, the catalyst bearing poly(aryl ether) wedges in the 6,6'-positions of BINOL gave the highest enantioselectivity (up to 87% ee).

The dendritic 2-amino-2'-hydroxy-1,1'-binaphthyl (NOBIN)-derived Schiff base ligands, displayed in Scheme 6, have been applied in the titanium-

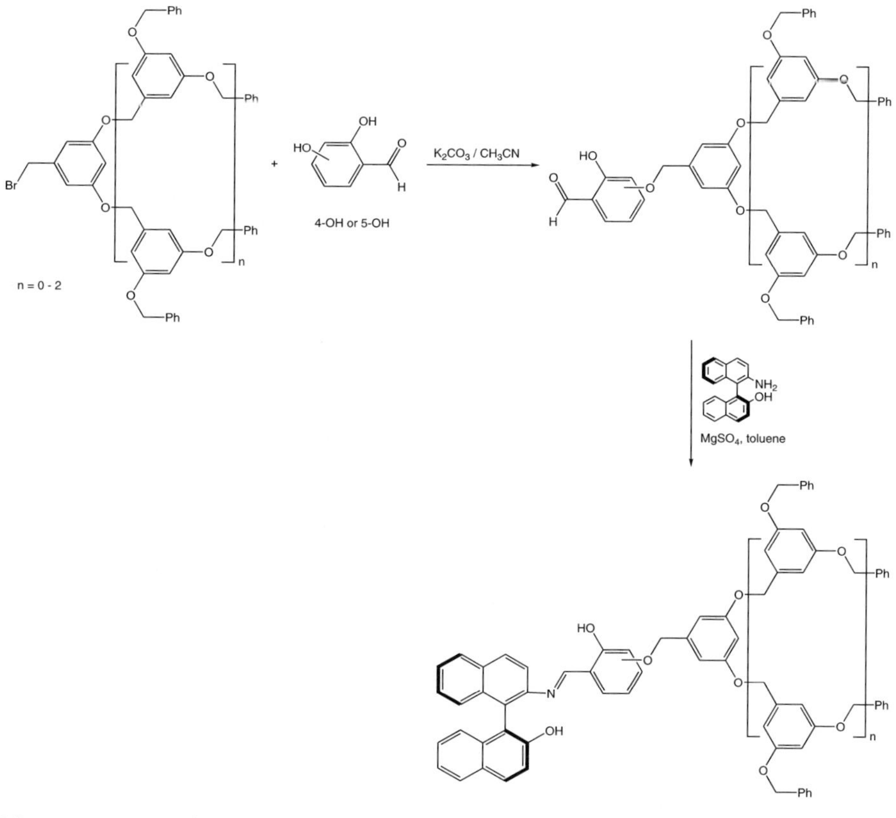

Scheme 6 NOBIN derivatives bearing dendritic poly(aryl ether) wedges. These systems have been studied in the titanium-catalyzed hetero-Diels–Alder reaction of Danishefsky's diene with aldehydes

catalyzed hetero-Diels–Alder reaction of Danishefsky's diene (1-methoxy-3-(trimethylsilyloxyl)buta-1,3-diene) with aldehydes [47]. These reactions afforded the corresponding 2-substituted 2,3-dihydro-4H-pyran-4-ones in quantitative yields and with excellent enantioselectivities (up to 97% ee). The disposition of the dendritic wedges and the size of the dendron in the ligands were found to have significant impact on the enantioselectivity of the catalytic reaction. The recovered catalysts could be reused without further addition of titanium reagent or a carboxylic acid additive for at least three cycles, retaining similar activity and enantioselectivity throughout the process. The other important observation has been the high degree of asymmetric amplification for the dendritic system.

4.2
Pyrphos-Based Catalysts Bearing Dendritic Wedges

In 2004, Chan et al. reported the synthesis of dendritic ligands bearing the chiral Pyrphos ligand at the focal point of Fréchet-type polyether dendrons (Fig. 17) [48]. The relationship between the primary structure of the dendrimer and its catalytic properties was established in the Rh-catalyzed hydrogenation of Z-methyl-α-acetamidocinnamate. For the systems containing the first- and second-generation dendritic wedges, high enantioselectivities (up to 98% ee) were observed. The third-generation catalyst gave lower enantioselectivity (95% ee) and a significantly decreased rate of conversion. Upon

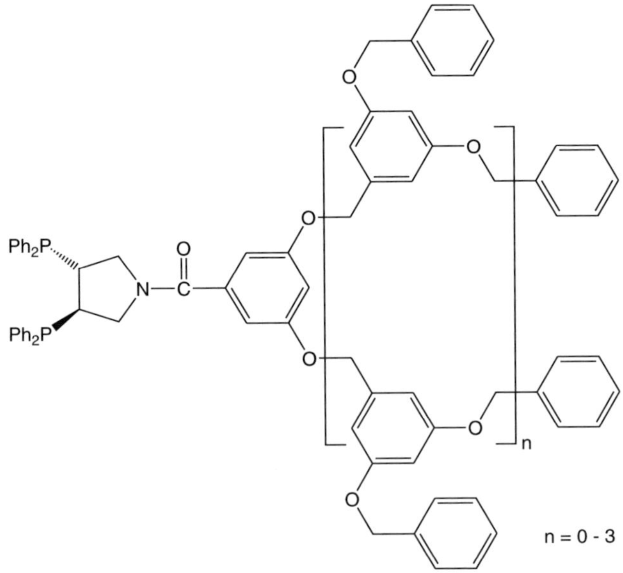

Fig. 17 A Pyrphos derivative bearing an N-bound poly(aryl ether) dendron

Fig. 18 Conversion curves for the Rh-catalyzed hydrogenation of *Z*-methyl-α-acetamido-cinnamate using the dendritic catalysts displayed in Fig. 17 [48]

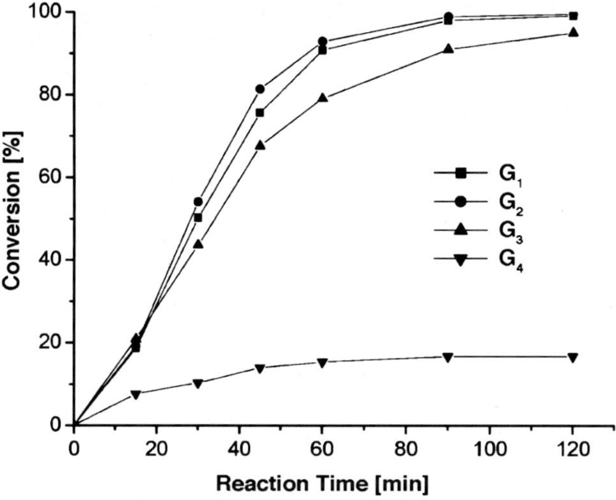

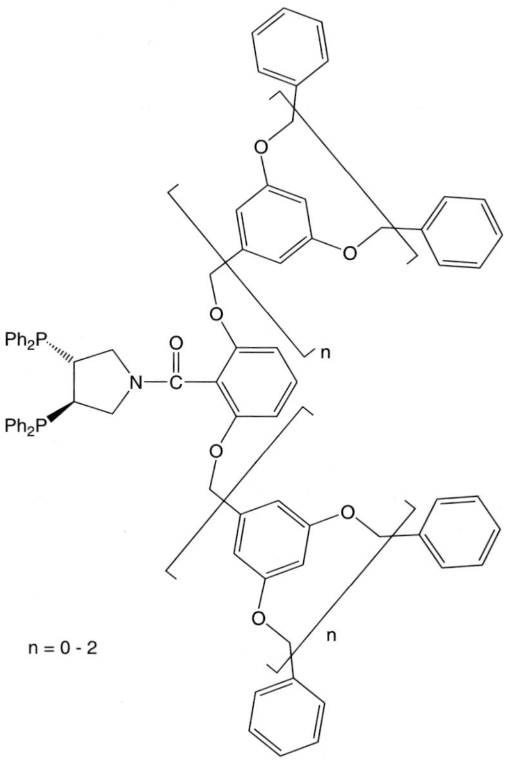

Fig. 19 Dendronized Pyrphos ligands bearing backfolded poly(aryl ether) dendrons

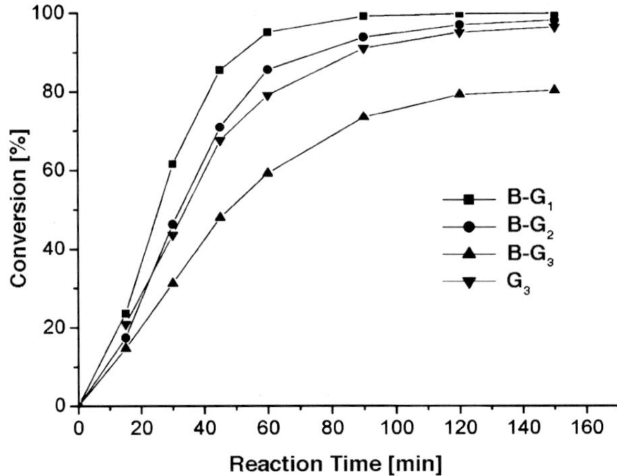

Fig. 20 Conversion curves for the Rh-catalyzed hydrogenation of Z-methyl-α-acetamido-cinnamate using ligands bearing the backfolded poly(aryl ether) dendrons displayed in Fig. 19. (B-G_x) compared with that of the third-generation catalyst displayed in Fig. 17 (G_3) [48]

going from generation 3 to 4, the dendritic catalyst almost completely lost its activity (Fig. 18).

This gradual decrease in reactivity is thought to be due to the increased steric shielding of the metal centre by the attached dendron on going to higher generations. To demonstrate this particular effect, Chan et al. designed "backfolded" dendrons by modifying the branching pattern of the dendritic wedges (Fig. 19). These backfolded linkages were expected to increase the degree of steric congestion around the catalytically active core. The conversion curves for the hydrogenations catalyzed by these backfolded dendrimers are represented in Fig. 20. Although the first-generation dendrimer displayed the same behaviour as the reference system in Fig. 17, the sterically demanding wedges of the higher-generation dendrimers significantly influenced the reactivity of these catalysts. This effect was most pronounced upon going from generation 2 to 3, a behaviour which is consistent with the effective encapsulation of the active core by the backfolded dendrimer. Furthermore, unlike the systems discussed above, the enantioselectivity of the Rh-catalyzed hydrogenation decreased for the higher dendrimer generations.

4.3
Dendrimer-Fixed Chiral Diamine-Based Catalysts

As for the exodendrally functionalized dendrimer catalysts (Sect. 3.2), chiral diamine ligands have also been the objects of study in the investigation into the catalytic behaviour of core-functionalized dendrons.

Deng et al. synthesized chiral diamine dendritic ligands using (*S,S*)-DPEN as ligand system and Frechét-type dendritic wedges as supporting material (Fig. 21) [49]. Asymmetric transfer hydrogenation reactions were studied using acetophenone as the model substrate. Compared with the monomeric Ru[(*S,S*)-DPEN] complex, a slightly enhanced reactivity was observed for the dendritic catalysts, as well as high enantioselectivities (> 96% ee).

Following up this work, the same group recently published the synthesis of "hybrid" dendritic ligands containing a combination of dendritic chiral DPEN and Fréchet polyether dendrons (Scheme 7) [50]. The solubility of these hybrid dendrimers was found to be controlled by the polyether dendron. Compared with the simple core-functionalized systems displayed in Fig. 21, the hybrid dendrimers showed similar catalytic activity but reduced recyclability.

Another possibility to immobilize the DPEN ligand is the functionalization of the phenyl rings, as exemplified in Fig. 22 [51–53]. The Ru-catalyzed hydrogenation of acetophenone was chosen as a test reaction, using 2-propanol and toluene in a 1 : 1 ratio as solvent system. The enantioselectivities of the first- and second-generation dendritic catalysts were comparable to that of Noyori's catalyst. However, the third-generation catalyst gave lower enantioselectivity (84%) and significantly decreased activity (45% conversion, 1 mol % catalyst) under the same conditions, even at a higher temperature (50 °C, 51% conversion and 88% ee). The sudden loss of activity for the third-generation dendritic catalyst may be thus attributed to the change in dendrimer conformation, from an extended to a more compact globular structure with increase of the steric requirements of the dendritic branches. This "encapsulation" of the active species by the dendrimer provides a barrier

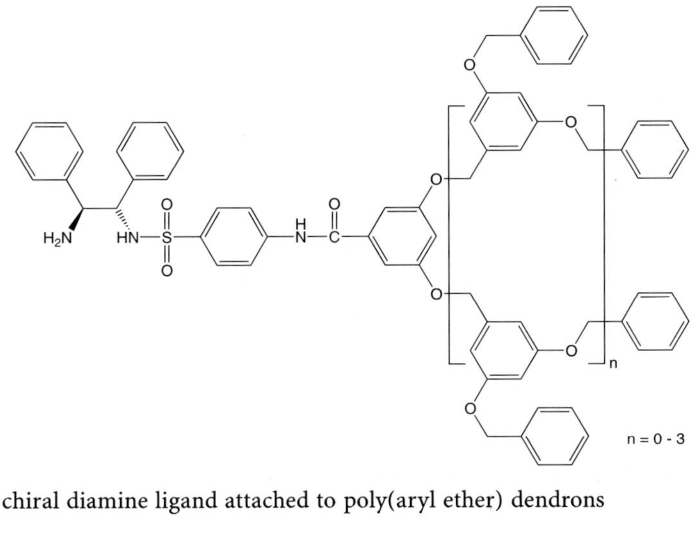

Fig. 21 A chiral diamine ligand attached to poly(aryl ether) dendrons

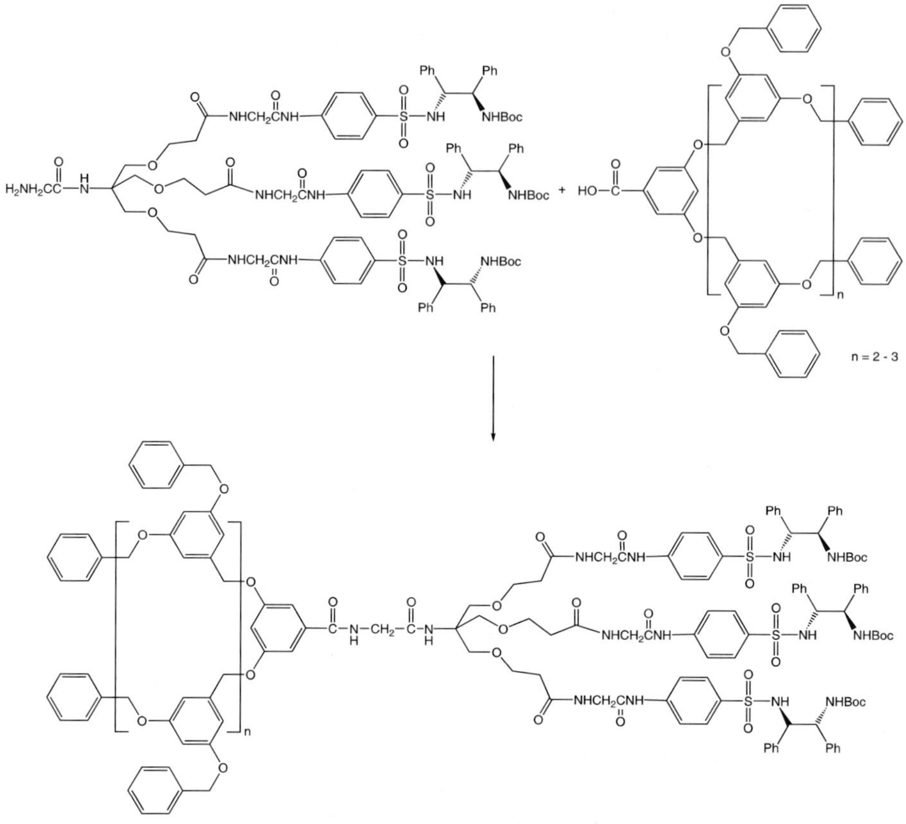

Scheme 7 "Hybrid" dendritic diamine ligands for Ru-catalyzed asymmetric transfer hydrogenation

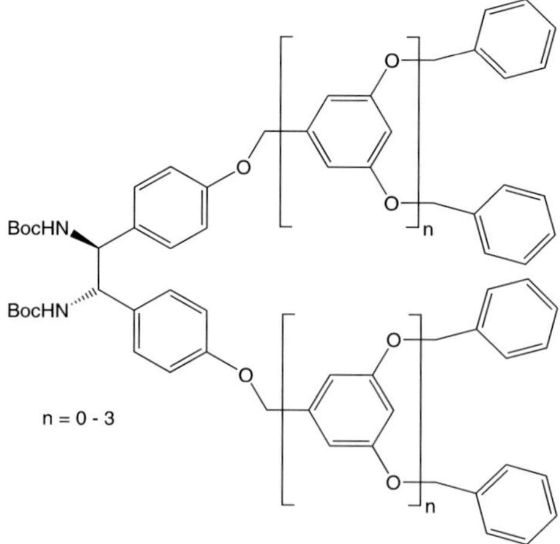

Fig. 22 Attachment of poly(aryl ether) dendrons to the phenyl substituents in the backbone of DPEN

to the diffusion of the substrate into the catalytically active core of the dendritic catalyst. Upon increase of the hydrogen pressure to 70 atm, complete conversion within 20 h was achieved.

4.4
Dendritic Bisoxazoline-Based Chiral Catalysts

In 2003, Chen and Fan synthesized a series of copper(II) complexes with chiral bis(oxazoline) ligands, which were disubstituted by Fréchet-type polyether dendrons at the carbon atom linking the two oxazolines (Fig. 23) [54]. These complexes were used as Lewis acid catalysts in enantioselective aldol reactions in aqueous media. High yields but only moderate enantioselectivities were obtained, which were comparable with those resulting from the corresponding small molecular catalysts. In fact, the dendritic substituents, which were assumed to affect the structure of the active site, did not decrease the enantioselectivity or the yield but gave slightly higher enantioselectivities and yields for the higher dendron generations.

Malmström and Moberg attached first- to fourth-generation dendritic substituents based on 2,2-bis(hydroxymethyl)propionic acid and (1R,2S,5R)-methoxyacetic acid to 2-(hydroxymethyl)pyridinooxazoline and bis-[4-(hydroxymethyl)oxazoline] compounds (Fig. 24) [55]. These new ligands were assessed in palladium-catalyzed allylic alkylations. The first type of ligands gave rise to enantioselectivities similar to that of a simple benzoyl ester derivative, whereas the latter type of ligands afforded products with higher selectivity than the analogous benzoyl ester. In general, the activity of the dendritic catalysts decreased with increasing generation.

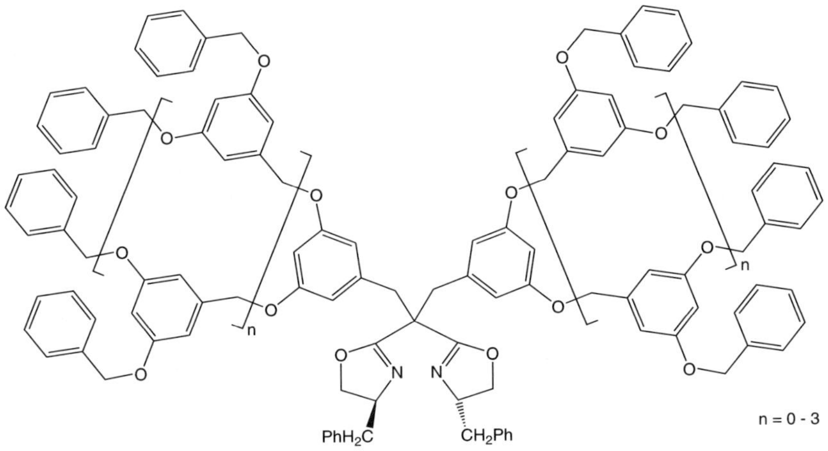

Fig. 23 The dendronized bisoxazolines studied by Chen and Fan

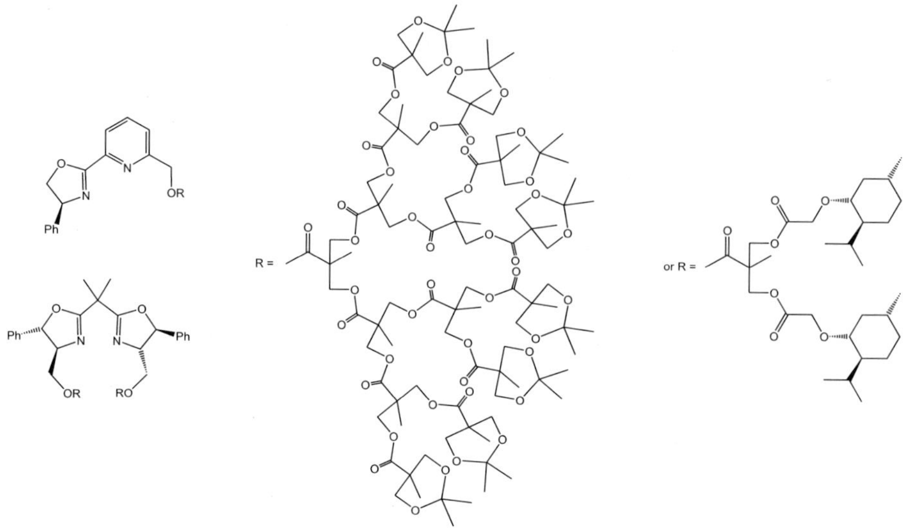

Fig. 24 The dendronized oxazoline ligands studied by Moberg et al.

4.5
Other Ligand Systems and Biphasic Catalysis

A series of chiral dendritic ligands derived from cinchonidine and Fréchet-type dendritic wedges up to generation 3 have been reported by van Koten et al. (Fig. 25) [56]. These dendritic ligands were tested as catalysts in the biphasic alkylation of *N*-(diphenylmethylene)glycine isopropyl ester with benzyl bromide (Scheme 8). The highest enantioselectivities (76% ee) were observed for the second-generation catalysts in aqueous 25% NaOH or aqueous 50% KOH as reaction media. Comparing the whole series of dendron

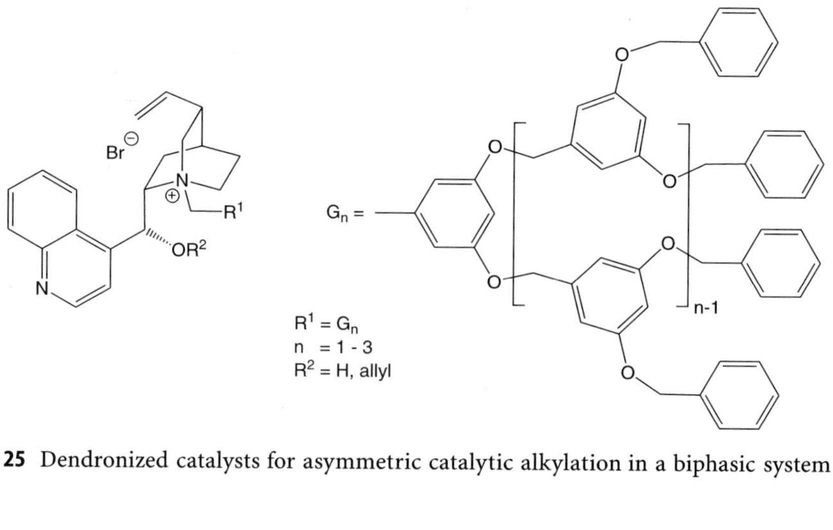

Fig. 25 Dendronized catalysts for asymmetric catalytic alkylation in a biphasic system

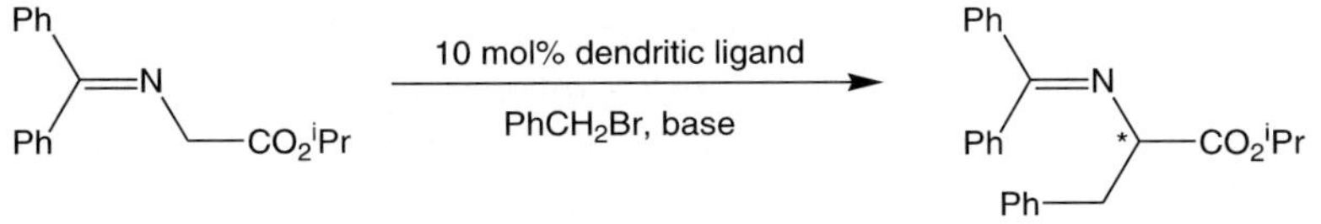

Scheme 8 Asymmetric alkylation of *N*-(diphenylmethylene)glycine isopropyl ester with benzyl bromide catalyzed by the dendronized catalysts shown in Fig. 25

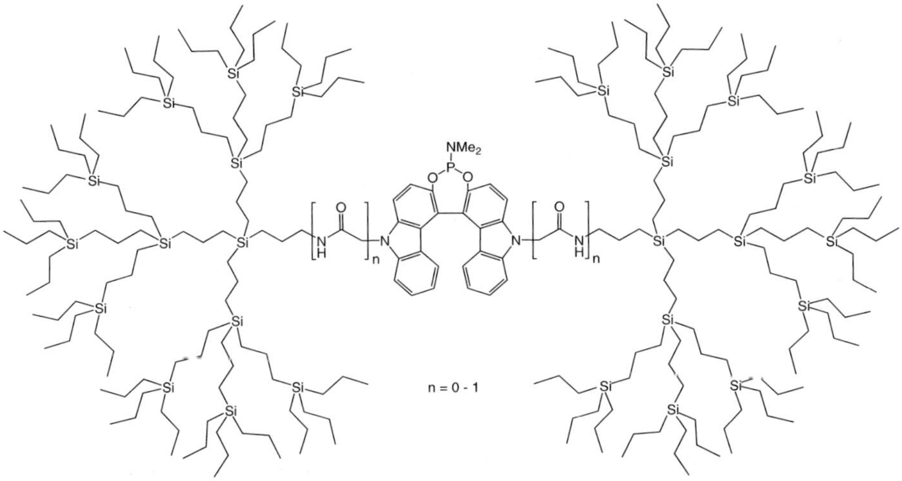

Fig. 26 A dendronized BICOL rhodium catalyst developed by Reek et al.

generations, it appears that the increase in steric hindrance due to the bulkiness of the dendritic wedge seems to have little effect on the enantioselectivity in this process.

Recently, Reek et al. published the synthesis of a $9H,9'H$-[4,4']bicarbazole-3,3'-diol (BICOL)-based chiral monodentate phosphoramidite ligand, which was functionalized with two different third-generation carbosilane dendritic wedges (Fig. 26) [57]. As reference reaction in the catalytic study, the rhodium-catalyzed asymmetric hydrogenation of Z-methyl-α-acetamido-cinnamate was chosen. Using a ligand-to-rhodium ratio of 2.2 led to enantio-selectivities which were comparable to the results obtained using the parent BINOL-derived monodentate phosphoramidite MonoPhos.

5
Polymer-Supported Chiral Dendritic Catalysts

Both the activity and the selectivity of heterogeneously immobilized molecular catalysts are frequently reduced with respect to the performance of their soluble analogues used under homogeneous conditions. The reasons for this are manifold, and include hindered diffusion processes or a significant change in the preferred conformations within the ligand shell of the catalytic moiety. A way out of this dilemma may be the use of dendronized polymers in which the catalysts are attached in a well-defined way to the dendritic sections, thus ensuring a well-defined microenvironment which is similar to that of the soluble molecular species or at least to the dendrimer catalysts themselves.

The combination of an efficient control over the environment of the active sites in a multi-functional catalyst and its immobilization within an insoluble macromolecular support was pioneered by Seebach et al. In their approach, the chiral ligand to be immobilized was placed in the core of a polymerizable dendrimer, followed by copolymerization of the latter with styrene as shown in Scheme 9 [58]. In this way, no further cross-linking agent was necessary, since the dendrimer itself acted as cross-linker. The dendritic branches are thought to act as spacer units, keeping the obstructing polystyrene backbone

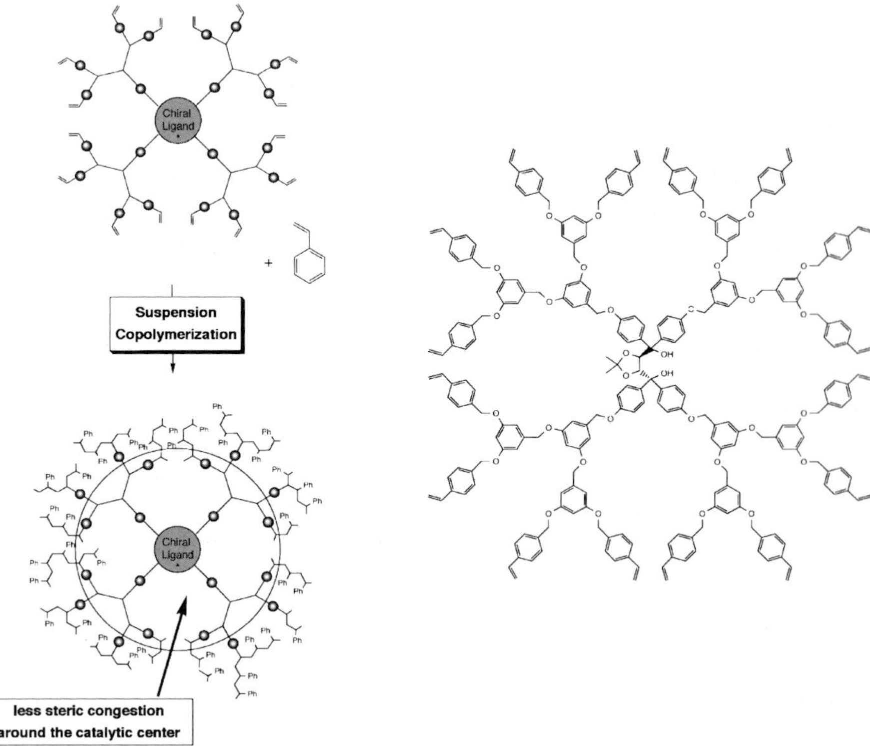

Scheme 9 Dendronized TADDOL ligands which are cross-linked to form a dendritic network [58]

away from the catalytic centres and leading to better accessibility, and thus to enhanced catalytic activity.

A series of styryl-substituted TADDOL derivatives with flexible, rigid or dendritically branching spacers between the TADDOL core and the styryl groups have been prepared and used as cross-linkers in styrene suspension polymerization, leading to polymer beads ca. 400 μm in diameter. These in turn were loaded with titanate, and used for the Lewis acid catalyzed addition of Et_2Zn to PhCHO as a test reaction. A comparison of the enantioselectivities and degrees of conversion (both up to 99%) showed that these polymer-incorporated Ti-TADDOLates were highly efficient catalysts for this process. The best performance over 20 cycles of the test reaction was obtained with the TADDOL bearing four first-generation Fréchet branches with eight peripheral styryl groups. The enantioselectivity, the rate of reaction and the swelling factor were essentially unchanged after numerous operations carried out with the catalytic beads and a degree of loading of 0.1 mmol TADDOLate/g polymer, with or without stirring. The rate of conversion for the dendritically polymer-embedded Ti-TADDOLate was greater than that found for the corresponding monomer.

Seebach et al. also prepared salen derivatives carrying two to eight styryl groups for cross-linking copolymerization with styrene [59]. The salen cores were derived either from (R,R)-diphenylethylenediamine or (R,R)-cyclohexanediamine, and the styryl groups were attached to the salicylic aldehyde moieties using Suzuki or Sonogashira cross-coupling reactions, as well as phenol etherification with dendritic styryl-substituted Fréchet-type benzylic branch bromides. Subsequent condensation with the diamines provided the chiral salens (Fig. 27). These polymer-bound Mn and Cr complexes were used as catalysts for the stereoselective epoxidation of phenyl-substituted olefins, as well as for catalyzed dihydropyranone formation from Danishefsky's diene and aldehydes.

There are several remarkable features of these immobilized salens, notably the fact that the dendritic branches do not appear to decrease the catalytic activity with respect to the complexes in solution. Moreover, the reactions with dendritic catalysts incorporated in polystyrene gave products of essentially the same enantiopurity as those observed in homogeneous solution, with the dendritically substituted or with the original Jacobsen–Katsuki complexes. Some of the Mn-loaded beads were stored for a year without loss of activity. Especially, the biphenyl- and acetylene-linked salen polymers gave Mn complexes of excellent performance, which after ten catalytic runs showed no loss of enantioselectivity or degree of conversion.

In another example of the dendronization of solid supports, Rhee et al. described the design of silica-supported chiral dendritic catalysts for the enantioselective addition of diethylzinc to benzaldehyde (Fig. 28) [60–62]. The immobilized dendritic systems were formed in two different ways: one by stepwise propagation of dendrimers and the other by direct immobilization

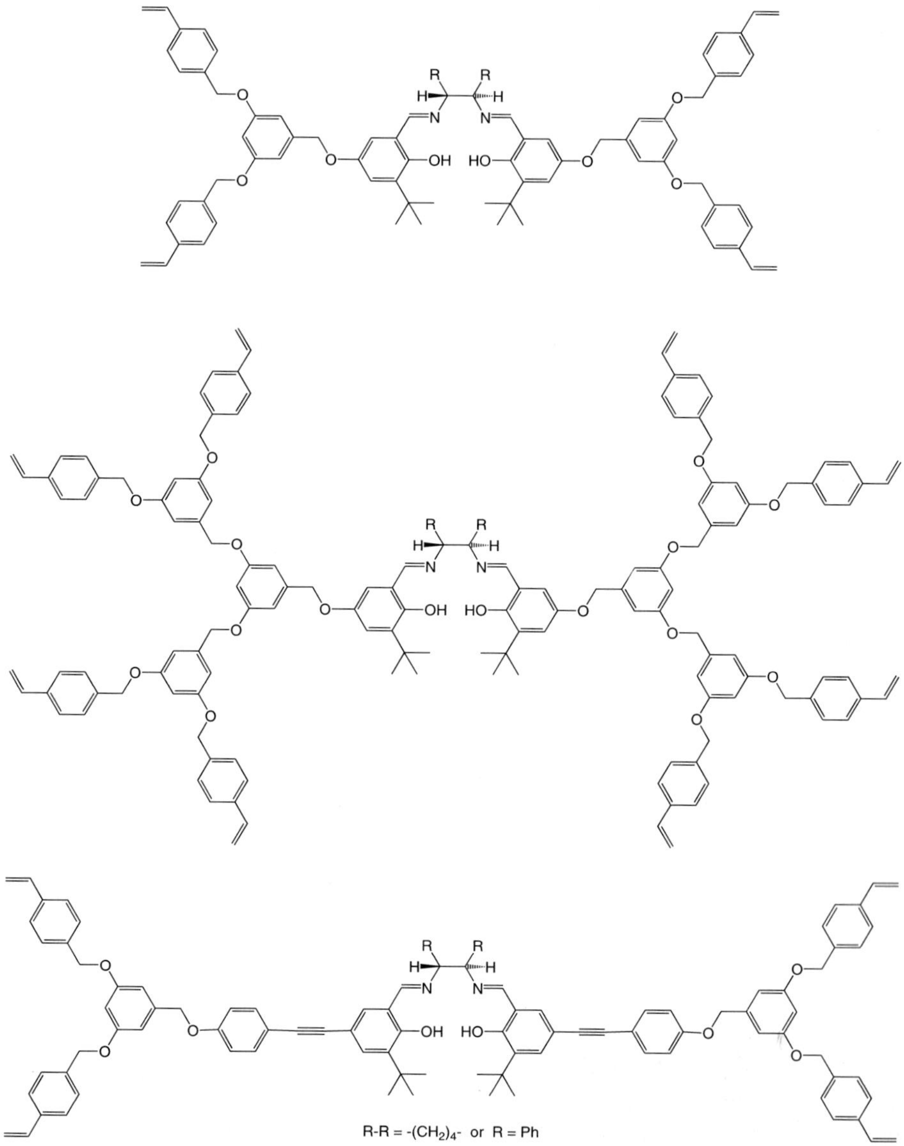

R-R = -(CH₂)₄- or R = Ph

Fig. 27 Dendronized salen ligands which were subsequently immobilized by cross-linking of the styryl units

of the complete dendrimers on silica. From the former alternative, i.e. the dendrimer catalyst preparation by stepwise propagation of dendrimers on silica, symmetric hyperbranching was found to be a prerequisite in order to suppress the unfavourable racemic reaction taking place on the naked surface. Moreover, the control of hyperbranching was important not only to retain the accessibility of the active sites, but also to relieve the multiple interactions between the chiral active sites. In the alternative method of dendrimer catalyst preparation by direct immobilization of preformed dendrimers on silica, the participation of surface silanol groups in the racemic reaction first had

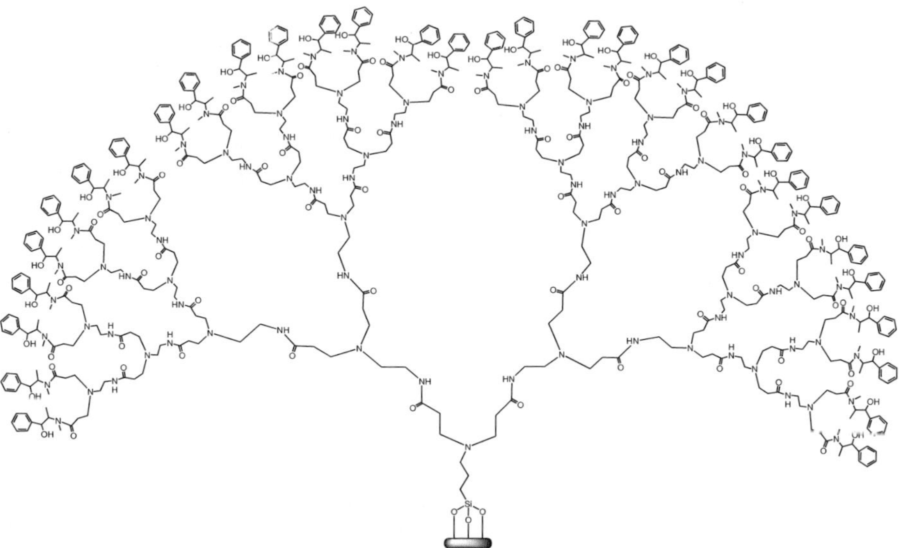

Fig. 28 Dendronization of a silica support

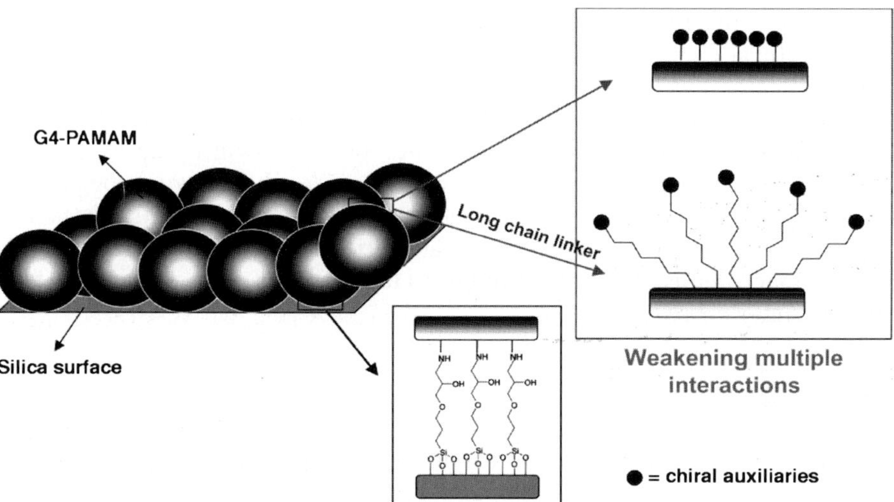

Fig. 29 Effect of the linker chain lengths on the intermolecular interaction of catalytic dendrimers immobilized on silica supports [62]

to be effectively suppressed by appropriate functionalization. The substitution of terminal end groups with a long alkyl chain was found to suppress the multiple interactions between the active sites (Fig. 29).

Finally, homopolymers of bis(oxazoline) ligands have been used to prepare efficient catalysts for cyclopropanation reactions. However, the reduced accessibility to most of the bis(oxazoline) moieties along with the high substitutional lability of copper(I)/(II) leads to a low degree of metal loading. As a consequence, the transmission of chiral information from the metallated polymer is inefficient. The use of suitable dendrimers as cross-linkers in the polymerization process allows a higher level of metallation.

In 2003, Mayoral et al. proved that by using this strategy, the productivity of chiral cyclopropanes per molecule of chiral ligands immobilized on a dendrimer greatly increased, which led to an improvement in the ligand economy and the chirality transfer [63].

6
Conclusion and Outlook

As has been emphasized at the beginning of this overview of asymmetric dendrimer catalysis, the kinetically controlled stereoselection depends on very small increments of free activation enthalpy. It is therefore an excellent sensitive probe for "dendrimer effects" and will continue to be studied in this fundamental context. As monodispersed macromolecules, chiral dendrimer catalysts provide ideal model systems for less regularly structured but commercially more viable supports such as hyperbranched polymers.

However, the results obtained in recent years have also established that the structural characteristics of the established dendrimer systems, such as the absence of a well-defined secondary structure, have limited the development of efficient abiotic enzyme mimics based on dendrimers. To achieve this ambitious goal, more efforts in dendrimer synthesis will be necessary. The use of dendritic catalysts in biphasic solvent systems has only just begun and appears to be a particularly fruitful field for further developments. These utilitarian aspects aside, it is the aesthetic attraction of these topologically highly regular macromolecules that continues to fascinate those working in the field of dendrimer catalysis.

References

1. van Heerbeek R, Kamer PCJ, van Leeuwen PWNM, Reek JNH (2002) Chem Rev 102:3717
2. Dijkstra HP, van Klink GPM, van Koten G (2002) Acc Chem Res 35:798
3. Knapen JWJ, van der Made AW, de Wilde JC, van Leeuwen PWNM, Wijkens P, Grove DM, van Koten G (1994) Nature 372:659

4. Gossage RA, van de Kuil LA, van Koten G (1998) Acc Chem Res 31:423
5. Brunner H, Altmann S (1994) Chem Ber 127:2285
6. Brunner H (1995) J Organomet Chem 500:39
7. Noyori R (2002) Angew Chem Int Ed 41:2008
8. Peerlings HW, Meier EW (1997) Chem Eur J 3:1563
9. Seebach D, Rheiner PB, Greiveldinger G, Butz T, Sellner H (1998) Top Curr Chem 197:125
10. de Groot D, de Waal BFM, Reek JNH, Schenning APHJ, Kamer PCJ, Meijer EW, van Leeuwen PWNM (2001) J Am Chem Soc 123:8453
11. Sato I, Shibata T, Ohtake K, Kodaka R, Hirokawa Y, Shirai N, Soai K (2000) Tetrahedron Lett 41:3123
12. Köllner C, Pugin B, Togni A (1998) J Am Chem Soc 120:10274
13. Schneider R, Köllner C, Weber I, Togni A (1999) Chem Commun 2415
14. Togni A, Dorta R, Köllner C, Pioda G (1998) Pure Appl Chem 70:1477
15. Köllner C, Togni A (2001) Can J Chem 79:1762
16. Engel GD, Gade LH (2002) Chem Eur J 8:4319
17. Ribourdouille Y, Engel GD, Richard-Plouet M, Gade LH (2003) Chem Commun 1228
18. Hayashi T (1993) In: Ojima I (ed) Catalytic asymmetric synthesis. Wiley, Weinheim, p 325
19. Trost BM, van Vranken DL (1996) Chem Rev 96:395
20. Helmchen G, Pfaltz A (2000) Acc Chem Res 33:336
21. Johannsen M, Jørgensen KA (1998) Chem Rev 98:1689
22. Heumann A (1998) Palladium-catalyzed allylic substitutions. In: Beller M, Bolm C (eds) Transition metals for organic synthesis. Wiley, Weinheim, p 251
23. Breinbauer R, Jacobsen EN (2000) Angew Chem Int Ed 39:3604
24. Suzuki T, Hirokawa Y, Ohtake K, Shibata T, Soai K (1997) Tetrahedron Asymmetry 8:4033
25. Sato I, Kodaka R, Shibata T, Hirokawa Y, Shirai N, Ohtake K, Soai K (2000) Tetrahedron Asymmetry 11:2271
26. Sato I, Kodaka R, Hosoi K, Soai K (2002) Tetrahedron Asymmetry 13:805
27. Sato I, Hosoi K, Kodaka R, Soai K (2002) Eur J Org Chem 3115
28. Soai K, Sato I (2003) C R Chimie 6:1097
29. Chen YC, Wu TF, Deng JG, Liu H, Cui X, Zhu J, Jiang YZ, Choi MCK, Chan ASC (2002) J Org Chem 67:5301
30. Arai T, Sekiguti T, Lizuka Y, Takizawa S, Sakamoto S, Yamaguchi K, Sasai H (2002) Tetrahedron Asymmetry 13:2083
31. Chow HF, Wan CW (2002) Helv Chim Acta 85:3444
32. Laurent R, Caminade AM, Majoral JP (2005) Tetrahedron Lett 46:6503
33. Schmitzer A, Perez E, Rico-Lattes I, Lattes A, Rosca S (1999) Langmuir 15:4397
34. Schmitzer A, Perez E, Rico-Lattes I, Lattes A (1999) Tetrahedron Lett 40:2947
35. Rico-Lattes I, Schmitzer A, Perez E, Lattes A (2001) Chirality 13:24
36. Schmitzer AR, Franceschi S, Perez E, Rico-Lattes I, Lattes A, Thion L, Erard M, Vidal C (2001) J Am Chem Soc 123:5956
37. Schmitzer A, Perez E, Rico-Lattes I, Lattes A (2003) Tetrahedron Asymmetry 14:3719
38. Seebach D, Marti RE, Hintermann T (1996) Helv Chim Acta 79:1710
39. Rheiner PB, Sellner H, Seebach D (1997) Helv Chim Acta 80:2027
40. Fan QH, Chen YM, Chen XM, Jiang DZ, Xi F, Chan ASC (2000) Chem Commun 789
41. Deng GJ, Fan QH, Chen XM, Liu GH, Chan ASC (2002) Chem Commun 1570
42. Deng GJ, Fan QH, Chen XM, Liu GH (2003) J Mol Catal A 193:21
43. Yu HB, Hu QS, Pu L (2000) Tetrahedron Lett 41:1681

44. Deng GJ, Yi B, Huang YY, Tang WJ, He YM, Fan QH (2004) Adv Synth Catal 346:1440
45. Fan QH, Liu GH, Chen XM, Deng GJ, Chan ASC (2001) Tetrahedron Asymmetry 12:1559
46. Liu GH, Tang WJ, Fan QH (2003) Tetrahedron 59:8603
47. Ji B, Yuan Y, Ding K, Meng J (2003) Chem Eur J 9:5989
48. Yi B, Fan QH, Deng GJ, Li YM, Qiu LQ, Chan ASC (2004) Org Lett 6:1361
49. Chen YC, Wu TF, Deng JG, Liu H, Jiang YZ, Choi MCK, Chan ASC (2001) Chem Commun 1488
50. Chen YC, Wu TF, Jiang L, Deng JG, Liu H, Zhu J, Jiang YZ (2005) J Org Chem 70:1006
51. Liu W, Cui X, Cun L, Wu J, Zhu J, Deng JG, Fan QH (2005) Synlett 1591
52. Liu W, Cui X, Cun L, Zhu J, Deng JG (2005) Tetrahedron Asymmetry 16:2525
53. Liu PN, Chen YC, Li X, Qiang T, Yong Q, Deng JG (2003) Tetrahedron Asymmetry 14:2481
54. Yang BY, Chen XM, Deng GJ, Zhang YL, Fan QH (2003) Tetrahedron Lett 44:3535
55. Malkoch M, Hallman K, Lutsenko S, Hult A, Malmström E, Moberg C (2002) J Org Chem 67:8197
56. Guillena G, Kreiter R, van de Coevering R, Klein Gebbink RJM, van Koten G, Mazon P, Chinchilla R, Najera C (2003) Tetrahedron Asymmetry 14:3705
57. Botman PNM, Amore A, van Heerbeek R, Back JW, Hiemstra H, Reek JNH, van Maarseveen JH (2004) Tetrahedron Lett 45:5999
58. Sellner H, Rheiner PB, Seebach D (2002) Helv Chim Acta 85:352
59. Sellner H, Karjalainen JK, Seebach D (2001) Chem Eur J 7:2873
60. Chung YM, Rhee HK (2002) Chem Commun 238
61. Chung YM, Rhee HK (2002) Catal Lett 82:249
62. Chung YM, Rhee HK (2003) C R Chimie 6:695
63. Diez-Barra E, Fraile JM, Garcia JI, Garcia-Verdugo E, Herrerias CI, Luis SV, Mayoral JA, Sanchez-Verdu P, Tolosa J (2003) Tetrahedron Asymmetry 14:773

Top Organomet Chem (2006) 20: 97–120
DOI 10.1007/3418_033
© Springer-Verlag Berlin Heidelberg 2006
Published online: 5 July 2006

Dendrimer-Encapsulated Bimetallic Nanoparticles: Synthesis, Characterization, and Applications to Homogeneous and Heterogeneous Catalysis

Bert D. Chandler (✉) · John D. Gilbertson

Department of Chemistry, Trinity University, 1 Trinity Place, San Antonio, TX 78212, USA
Bert.chandler@trinity.edu

Abstract We review the preparation, characterization, and properties of dendrimer-templated bimetallic nanoparticles. Polyamidoamine (PAMAM) dendrimers can be used to template and stabilize a wide variety of mono- and bimetallic nanoparticles. Depending on the specific requirements of the metal system, a variety of synthetic methodologies are available for preparing nanoparticles with diameters on the order of 1–3 nm with narrow particle size distributions. The resulting dendrimer-encapsulated nanoparticles, or DENs, have been physically characterized with electron microscopy techniques, as well as UV-visible and X-ray photoelectron spectroscopies.

For certain metal systems, the chemical properties of bimetallic DENs include selective extraction from the dendrimer interior into organic solvents. Catalytic properties include homogeneous hydrogenation catalysis; heterogeneous hydrogenation and oxidation catalysis have also been examined. Homogeneous hydrogenation studies indicate that synergism in catalytic activity often occurs when two metals are intimately mixed in nanoparticles. DENs can also be deposited onto a variety of solid substrates and the organic dendrimer template thermally removed. The resulting activated nanoparticles are also active catalysts, and have been further characterized with infrared spectroscopy of adsorbed CO. Relationships between these heterogenized systems and the solution DENs are also discussed.

Keywords Heterogeneous catalysis · Homogeneous catalysis · Nanoparticle synthesis · PAMAM dendrimers · Transmission electron microscopy · UV-visible spectroscopy · X-ray photoelectron spectroscopy

Abbreviations

PAMAM	polyamidoamine
PPI	polypropyleneimine
DENs	dendrimer-encapsulated nanoparticles
Gx	generation X (4, 5, ...)
HRTEM	high resolution transmission electron microscopy
EDS	energy dispersive spectroscopy
XPS	X-ray photoelectron spectroscopy
MPC	monolayer protected cluster
SAM	self-assembled monolayer
TOF	turnover frequency
DRIFTS	diffuse reflectance infrared Fourier transform spectroscopy
AA	atomic absorption spectroscopy
ICP-MS	inductively coupled plasma mass spectrometry
1,3-COD	1,3-cyclooctadiene

1
Dendrimers and Dendrimer-Encapsulated Nanoparticles

Dendrimers are hyperbranched polymers that emanate from a single core and ramify outward with each subsequent branching unit [1, 2]. In the commonly employed "divergent" synthesis, dendrimers can be prepared through sequential, alternating reactions of two smaller units, one of which has a point of bifurcation. As is described elsewhere in this volume, several classes of dendrimers are known, including polypropyleneimine (PPI), polyamidoamine (PAMAM), and Fréchet-type polyether dendrimers [1, 2].

Starburst PAMAM dendrimers (Fig. 1), a specific class of commercially available dendrimers that have repeating amine/amide branching units, have drawn considerable interest in recent years due to their potential applications in medicine, nanotechnology, and catalysis [3–7]. These dendrimers are readily functionalized to terminate in diverse moieties such as primary amines, carboxylates, hydroxyls, or hydrophobic alkyl chains. Because dendrimer size and end groups can be varied, they are typically named by their generation (G1, G2, etc.) and exterior functionality ($-NH_2$, $-OH$).

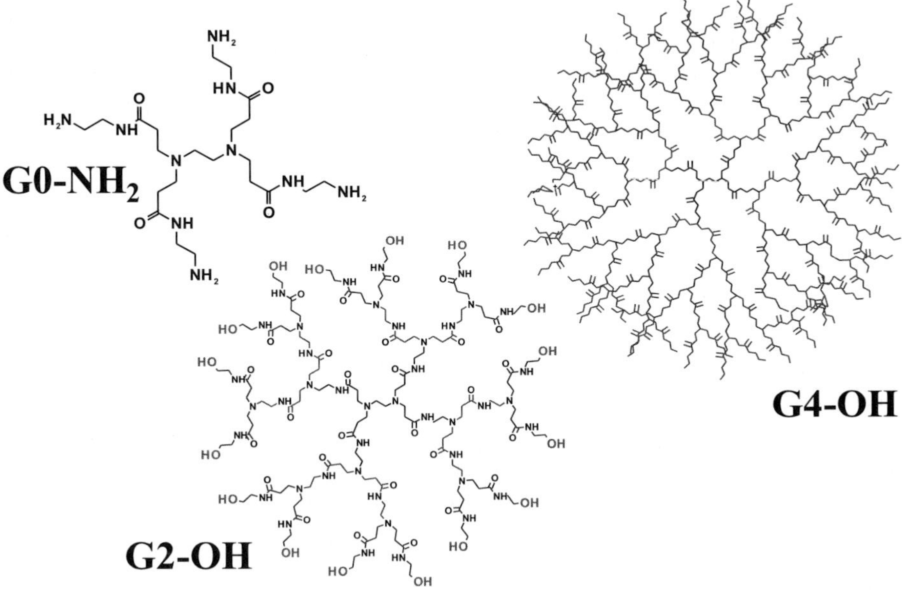

Fig. 1 Polyamidoamine dendrimers. The heteroatoms omitted from G4 – OH for clarity

Higher generation dendrimers (G4 and larger) adopt roughly spherical or globular structures with exterior branches becoming increasingly intertwined. The interweaving of dendrimer branches is distinct from the crosslinked branches that are found in many polymeric colloid stabilizers. The dendritic macromolecular architecture gives rise to a relatively open interior pocket for high generation dendrimers, while maintaining a closed, but porous exterior. The presence of these open spaces within the dendrimer interior and the synthetic control over their composition, architecture, and interior/exterior functionalities creates an environment that facilitates trapping guest species [8, 9].

PAMAM dendrimers can be used to template and stabilize reduced (zero valent) metal nanoparticles in solution [6, 10, 11]. Synthetic strategies for dendrimer-encapsulated nanoparticles (DENs) will be discussed in detail in the next section, but they are generally analogous to "ship in a bottle" syntheses. In the first step, metal ions (e.g. Cu^{2+}, Pd^{2+}, Pt^{2+}) are intercalated through the porous exterior of a PAMAM dendrimer and complexed to the interior amine groups. When electrons are added via a reducing agent (e.g. BH_4^-), reduced metal atoms, which are effectively trapped within the interior cavity, coalesce into a nanoparticle. Hence, the dendrimer plays multiple roles, serving both as a template for the number (and type) of metal atoms, and as a colloid stabilizer, preventing agglomeration after the particles form.

DENs are typically named by the dendrimer from which they are prepared along with the metal:dendrimer stoichiometry set in the initial syntheses, e.g. G5 – OH(Pt$_{50}$). With careful synthetic techniques, nanoparticles with very narrow particle size distributions (1.3 ± 0.3 nm) can be selectively prepared inside dendrimer interior cavities. Provided that metal reoxidation is prevented, DENs are stable for long periods of time and do not agglomerate, since the nanoparticles are trapped within the dendrimer framework [12–14].

2
Synthesis

Since the first report of dendrimer-encapsulated Cu nanoparticles [15], several types of mono and bi-metallic DENs have been prepared. DEN synthesis has been recently reviewed [9, 16], so only the synthesis of bimetallic DENs is described here. Bimetallic DENs can be prepared by one of three methods: co-complexation of metal salts, galvanic displacement, and sequential reduction. Several bimetallic systems have already been prepared inside PAMAM dendrimers; Table 1 summarizes the current literature and synthetic methods employed.

Table 1 Bimetallic DEN systems: synthetic methods and catalytic reactions

DEN	Synthesis[a]	Dendrimer(s)	Catalysis	Refs.
PdPt	Co-complex	G4 – OH	allyl alcohol hydrogenation	[19]
PdPt	Co-complex	G4 – OH	1,3-COD hydrogenation	[20]
PdRh	Co-complex	G4 – OH	1,3-COD hydrogenation	[22]
PdAu	Co-complex	G6 – Q116	allyl alcohol hydrogenation	[21]
PdAu	Co-complex	G4 – NH$_2$	CO oxidation[b]	[52]
PtAu	Galvanic	G5 – OH	CO oxidation[b]	[24]
PtCu	Co-complex	G5 – OH	CO oxidation[b] toluene hydrogenation[b]	[23]
PdAg	Co-complex	G4 – NH$_2$		[28]
AuAg	Seq. Red.	G3 – NH$_2$, G3.5 – NH$_2$, G5 – NH$_2$, G5.5 – NH$_2$	p-nitrophenol reduction	[30]
AuAg	Seq. Red.	G6 – OH, G8 – OH		[31]
[Au](Pd)	Seq. Red.	G6 – Q116	allyl alcohol hydrogenation	[21]
[Pd](Au)	Seq. Red.	G6 – OH		[21]
[Au](Ag)	Seq. Red.	G6 – OH		[31]
[AuAg](Au)	Seq. Red.	G6 – OH		[31]

[a] Co-complex = Co-complexation; Galvanic = Galvanic Displacement;
[b] Heterogeneous catalysis   Seq. Red. = Sequential Reduction

Generation 4 or higher hydroxyl-terminated PAMAM dendrimers are the most commonly employed precursors in DEN synthesis. The terminal hydroxyl groups avoid competitive binding of the metal ion(s) to the periphery amines of $Gx - NH_2$, which can lead to inter-dendrimer nanoparticles (particles outside of dendrimer cavities and stabilized by several dendrimers). Crooks' group has also developed PAMAM dendrimers with quaternary ammonium salts on their periphery (e.g. G6 – Q116) [17]. These dendrimers have proven extremely useful in preparing both mono- and bimetallic DENs as the net positive charges on the dendrimer surfaces help to prevent dendrimers from agglomerating in solution while maintaining water solubility [16].

2.1
Co-complexation

The most straightforward synthesis is commonly referred to as the co-complexation method. As Scheme 1 shows, the first step in the synthesis involves complexation of the metal ion(s) with the PAMAM dendrimer interior tertiary amine groups. Once the metal ion complexation reaction is complete, the metal ions (M_A^{n+} and M_B^{n+}) are reduced to nanoparticles ($M_A^0 M_B^0$) through the addition of a reducing agent such as $NaBH_4$. Metal:dendrimer stoichiometries are typically expressed as a ratio of metal atoms to interior amine groups. This is clearly an oversimplification as metal-dendrimer binding varies with different metal precursors and can involve interior amide bonds [15, 18]. Few detailed investigations of metal ion:dendrimer complexes and stoichiometries have been reported in the literature. This remains a topic of fundamental interest, particularly as it relates to the properties of the nanoparticles ultimately produced.

Co-complexation has been used to prepare a variety of bimetallic DENs, including PdPt [19, 20], PdAu [21], and PdRh [22] (Table 1). In a typical study, for example the PdPt system studied by Scott, Datye, and Crooks [19], K_2PdCl_4 and K_2PtCl_4 are added to a dilute aqueous solution of G4 – OH in a fixed metal-ion to dendrimer ratio of 40 : 1. The solution is stirred for four days to allow complete complexation of the metal ions with the interior ter-

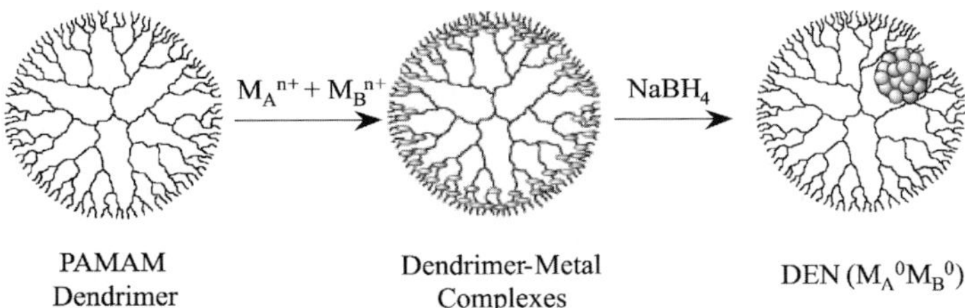

Scheme 1 Co-complexation synthesis

tiary amines. After complexation, 20-fold excess of BH_4^- was then added to reduce $G4 - OH[(Pd^{2+})_x(Pt^{2+})_{40-x}]$ to $G4 - OH[(Pd)_x(Pt)_{40-x}]$.

The relative complexation rates for different precursors are an important consideration for this synthetic scheme. In the Pt-Pd case, $PdCl_4^{2-}$ complexes quickly, while $PtCl_4^{2-}$ requires several days to completely react [18]. Given the great differences in these complexation rates, in practice, the synthesis described above involves sequential complexation of the ions. When metal precursors are introduced simultaneously to the dendrimer, the more reactive complex initially binds to the most accessible amine groups. It is also possible to control this process by adding the less reactive complex first, allowing binding to complete, and subsequently adding the second complex. This method was employed for PtCu DENs, allowing $PtCl_4^{2-}$ to bind $G5 - OH$ for 2 days before adding $Cu(NO_3)_2$. Elemental analysis of the resulting nanoparticles showed them to be consistently enriched in Cu, suggesting that either Pt(II) complexation was incomplete or that Cu(II) may displace some of the Pt(II)-dendrimer complexes [23].

2.2
Galvanic Displacement

Bimetallic DENs can also be prepared using the galvanic displacement method. This approach, shown in Scheme 2, utilizes an oxidizable DEN as a sacrificial reducing agent to prepare particles of more noble metals. Although it has been used to prepare a series of monometallic nanoparticles [9, 16], Cu displacement has only been utilized in the preparation of PtAu bimetallic DENs [24]. In this, Cu DENs are first prepared in hydroxyl-terminated dendrimers under anaerobic conditions. Oxygen-free solutions of K_2PtCl_4 and $HAuCl_4$ are then prepared, mixed and added to the reduced Cu solution. The Cu^0 DENs then act as the reducing agent for both Pt and Au preparing bimetallic nanoparticles through an intradendrimer redox displacement reaction [24]. The metal ratios reported through Cu displacement are consistent with ratios set in the initial syntheses, but this synthetic scheme has a general drawback of preparing particles with a somewhat wider size distribution than other methods. The Cu displacement synthesis is likely to

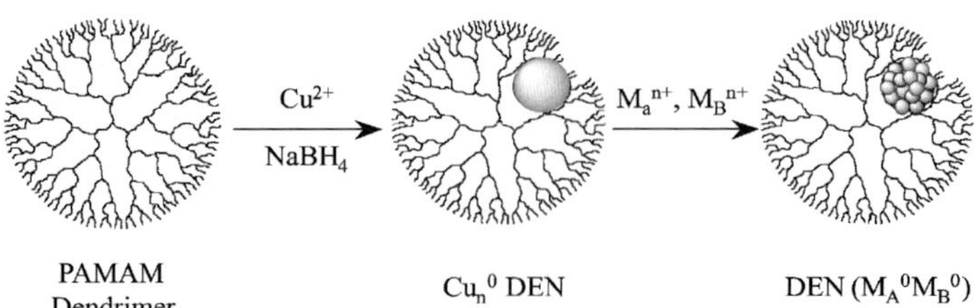

Scheme 2 Galvanic displacement synthesis

be most useful for rapid screening of metal ratios as it is a much faster route to Pt-based systems, leading to nanoparticles in several hours rather than several days.

Unless specific synthetic steps are undertaken to prepare well-defined morphologies (see below) bimetallic DENs prepared by co-complexation and galvanic displacement are generally described as "well-mixed" bimetallic nanoparticles or "alloy-type" particles. The critical aspect of these descriptions is the random, intimate mixing of the two metals within individual nanoparticles. We prefer the well-mixed term as the word alloy also refers to thermodynamically stable bulk solid solutions. In many cases, including nanoparticles prepared with dendrimers [24], bimetallic nanoparticles can be prepared throughout bulk immiscibility gaps [25, 26]. The well-mixed particle terminology avoids potential misconceptions associated with differing interpretations of alloy.

2.3
Sequential Reduction

The third method developed for preparing DENs is known as the sequential reduction method. This method, shown in Scheme 3 involves the initial complexation and reduction of a "seed" metal (M_A), followed by the complexation and subsequent reduction of the second metal (M_B) to produce the $M_A M_B$ system. The synthetic utility of this method is that it provides the means to access both well mixed and core/shell-type DENs. When the reducing agent used for M_B is a mild reductant, such as H_2 or ascorbic acid, core/shell nanoparticles with an M_A core and M_B shell can be selectively pre-

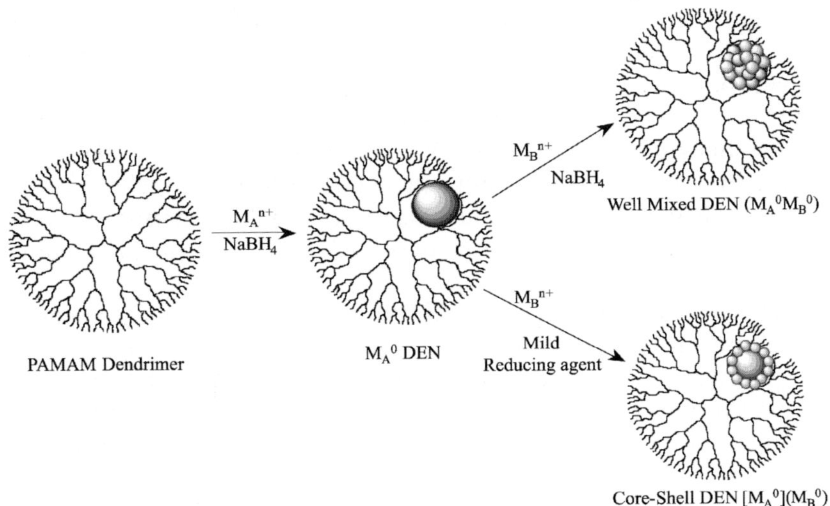

Scheme 3 Sequential reduction synthesis

pared. In the nomenclature for these core-shell nanoparticles, (M_A) denotes the core metal(s) and (M_B) indicates the exterior metal shell.

As Table 1 shows, sequential reduction can be used to selectively prepare either well-mixed or core/shell DENs [27–31]. In a typical synthesis, such as the preparation of [Au](Pd) nanoparticles [21], K_2PdCl_4 is added to a solution of $G6 - Q116(Au_{55})$ seeds and stirred to produce $G6 - Q116[Au_{55}](Pd^{2+})_{95}$. H_2 is then bubbled through the solution (10-fold excess of ascorbic acid was also explored as the reductant) to produce $G6 - Q116[Au_{55}](Pd)_{95}$.

3
Characterization of DENs

3.1
Electron Microscopy

Characterization of bimetallic DENs can be roughly separated into three separate categories: size, composition, and properties. Particle size is the most straightforward of these, and is best determined by High Resolution Transmission Electron Microscopy (HRTEM). Figure 2 shows a representative HRTEM micrograph from $G6 - Q116(Pd_{75}Au_{75})$ DENs. The particle size and distribution shown in Fig. 2 is typical for DENs, although the metal:dendrimer ratio and dendrimer generation used in the synthesis ultimately control these distributions. Most of the bimetallic systems in Table 1 have particle sizes in the range of 1–3 nm with standard deviation distributions of 0.5–1 nm.

Although alternate methods of determining particle sizes have not been explored with bimetallic DENs, a study by Kim, Garcia-Martinez, and Crooks is worth noting. In this study, the investigators used differential pulse voltammetry to estimate the size of dendrimer-templated Pd and Au nanoparticles [32]. This study estimated particle sizes to be very close to ideal sizes calculated from the metal:dendrimer stoichiometry and the metallic radius of each metal. Further, the study concluded that TEM measurements overestimated the size of the smallest Pd nanoparticles due to inadequate point-to-point resolution [32].

Particle composition is far more difficult to evaluate. Bulk elemental analysis [atomic absorption spectroscopy (AA) or inductively coupled plasma mass spectrometry (ICP-MS) are most common for metals] is useful in confirming the overall bimetallic composition of the sample, but provides no information regarding individual particles. Microscopy techniques, particularly Energy Dispersive Spectroscopy (EDS), has supported the assertion that bimetallic DENs are bimetallic nanoparticles, rather than a physical mixture of monometallics [16]. Provided the particle density is low

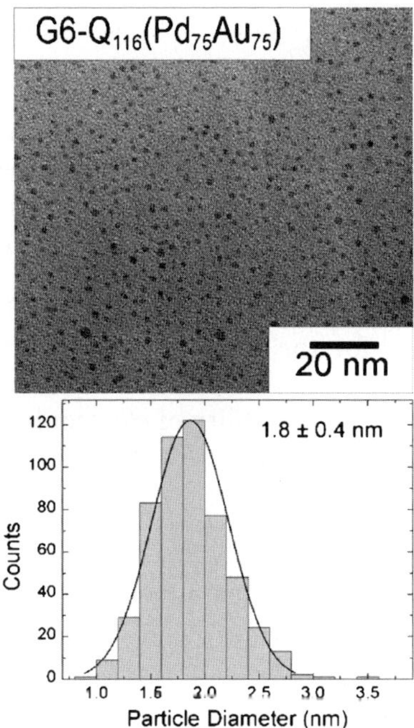

Fig. 2 HRTEM image of $G6-Q116(Pd_{75}Au_{75})$ DENs formed by the co-complexation method and a histogram of the particle-size distribution. Reprinted with permission from J Am Chem Soc, 2004, 126, 15583–15591. Copyright 2004 American Chemical Society

enough, it is possible to focus the electron beam to evaluate individual particles. These single particle EDS studies of $G4-OH(Pd_{30}Pt_{10})$ DENs [19] and activated dendrimer-templated PtAu nanoparticles [24] show both systems to be composed of two metals intimately mixed within individual nanoparticles.

Just as DENs particle sizes have some distribution (albeit relatively narrow), there is surely some distribution in particle compositions for bimetallic DENs. This is a fundamentally important aspect of DENs, particularly with regard to their catalytic properties; however, there are presently no reliable characterization methods for evaluating particle composition distributions. One method that has been applied to PdAu [21] and PtPd [19] DENs, as well as dendrimer-templated PtAu [24] is to collect single particle EDS spectra from several (15–20) nanoparticles. These experiments indicate that individual particle composition distributions may vary widely, but the difficulty in obtaining data from the smallest particles may skew the results somewhat. EDS spectra collected over large areas, which sample tens or hundreds of particles, generally agree well with the bulk composition measurements [24] and with stoichiometries set in nanoparticle synthesis [19, 21, 24].

3.2
UV-Visible Spectroscopy

UV-visible spectra of nanoparticles arise from two sources. The first, more general source is simple Rayleigh scattering that gives rise to the monotonic increase in absorption as wavelength decreases [33]. Au and Ag nanoparticles have intense surface plasmon bands that are valuable additional spectroscopic tools [33–35]. These bands, which arise from a concerted oscillation of nanoparticle electrons, shift with particle size and composition, and are therefore useful handles for the physical characterization of nanoparticle composition.

A representative sample of the utility of electronic spectra is illustrated in Fig. 3, which show a UV-vis study of PdAu DENs. Figure 3a shows changes in

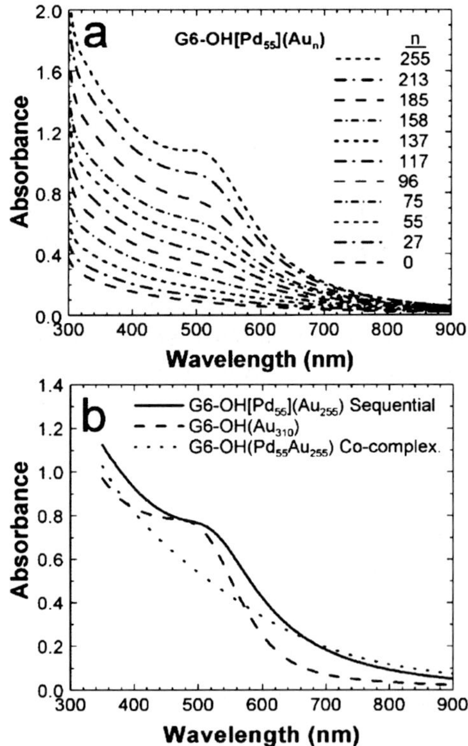

Fig. 3 UV-vis spectra of **a** 2.0 μM G6 – OH[Pd$_{55}$](Au$_n$) solutions (n = 0–255) prepared by the sequential-loading method. The DENs were prepared using ascorbic acid as the reducing agent. **b** 0.8 μM solutions of G6 – OH[Pd$_{55}$](Au$_{255}$), prepared by the sequential-loading method, and G6 – OH(Au$_{310}$) and G6 – OH(Pd$_{55}$Au$_{255}$) prepared by the co-complexation method. Reprinted with permission from J Am Chem Soc, 2004, 126, 15583–15591. Copyright 2004 American Chemical Society

electronic spectra of 2.0 µM solutions of core-shell $G6-OH[Pd_{55}](Au_n)$ as n increases from 0 to 255. As the Au content increases the plasmon shoulder develops at ~ 520 nm. This plasmon shoulder is typical of Au nanoparticles [33, 35] and suggests the presence of a Au shell around the Pd core. Figure 3b shows the spectra of $G6-OH[Pd_{55}](Au_{255})$ compared to spectra of $G6-OH(Au_{310})$ and the co-complexation product $G6-OH(Pd_{55}Au_{255})$. The plasmon band of the solution containing $G6-OH[Pd_{55}](Au_{255})$ is shifted to lower energy by ~ 10 nm compared to that of $G6-OH(Au_{310})$. Since both DENs are of comparable size, this shift in the plasmon band is indicative of structural differences between the two. More notable is the absence of the plasmon band in the spectrum of the solution containing the well-mixed $G6-OH(Pd_{55}Au_{255})$ DENs. Only monotonically increasing absorbance toward higher energy is observed [33]. A similar finding was reported for well-mixed PtAu DENs [24], indicating that the structural differences between core/shell and well-mixed morphologies are indeed reflected in UV-visible spectra.

3.3
X-Ray Photoelectron Spectroscopy

In conjunction with the UV-vis data, X-ray Photoelectron Spectroscopy (XPS) provides evidence for the complete reduction of the metal ions to nanoparticles. Studies from Rhee's group, which explored the degree of reduction in the $G4-OH(Pd^{2+}/Pt^{2+})$ system, provide a representative sample of XPS data [20]. In this study, the Pd/Pt ratio was varied while maintaining a constant metal : dendrimer ratio of 55 : 1. The peaks corresponding to $Pd(3d_{5/2})$ and $Pd(3d_{3/2})$ at 337.6 and 342.7 eV, respectively, were assigned to Pd^{2+}. After reduction, these peaks shift to 334.9 and 340.5 eV, respectively, and are consistent with Pd^0 [36]. Comparable shifts were observed for the Pt. The Pt^{2+} peaks at 72.5 eV $[Pt(4f_{7/2})]$ and 75.7 eV $[Pt(4f_{5/2})]$ shift to 71.3 and 74.4 eV, respectively, upon reduction and are consistent with Pt^0. Peaks for unreduced Pd and Pt were not reported, suggesting complete reduction of both metals occurred.

4
Nanoparticle Extraction from Dendrimer Interiors

Despite the utility of the physical methods described above, characterization of entities on the nanometer scale is still a problem. Well-mixed nanoparticles are not necessarily completely homogeneous, and one metal may preferentially segregate to the nanoparticle surface. Subtle differences in surface stoichiometries are presently extremely difficult to quantitatively evaluate with spectroscopy, even when the metal of interest has an intense surface plasmon.

The chemical reactivity of nanoparticle surfaces, presents interesting additional opportunities for evaluating nanoparticle surface composition. Some noble metal particles (Pd and Au in particular) can be extracted from the PAMAM dendrimer interiors into organic solution with long-chain thiols [37]. The resulting nanoparticles, referred to as Monolayer Protected Clusters (MPCs), retain the size distributions and spectroscopic characteristics of the original DENs and allow for recycling the expensive dendrimer [16].

This general extraction scheme can be extended to the selective extractions with a variety of Au and Ag DENs [31]. Using a combination of long-chain thiols and carboxylic acids, Crooks' group has shown it is possible to selectively extract Au and Ag monometallic DENs from mixtures of the two, and hence provided important chemical information about the shell of the nanoparticle [38]. In the case of the bimetallic DENs, selective extraction is a potentially powerful tool for characterizing particle surfaces, especially in the case of the core/shell DENs.

As Scheme 4 shows, it is possible to extract $[Au_{55}](Ag_n)$ nanoparticles (where $n = 95, 254, 450$) from an aqueous solution of $G6 - OH[Au_{55}](Ag_n)$ into hexane using n-dodecanethiol. The presence of a strong reducing agent, typically excess BH_4^-, is required to ensure the Ag shell is fully reduced (surface Ag atoms rapidly oxidize and the oxide layer does not bind to the thiol). This produces the MPC, which is described as $MPC - RSH[Au_{55}](Ag_n)$. Similar to the thiol chemistry, in the absence of a reducing agent, n-undecanoic acid will extract bimetallic nanoparticles that have Ag shells, but will not extract Au-shelled nanoparticles. This study was based on the strong literature precedent that n-alkanoicacids form self-assembled monolayers (SAMs) with Ag, provided there is an oxide layer present [39, 40].

This general scheme can also be extended to mixed metal cores to show that the shell composition is the key parameter for extraction. Using a se-

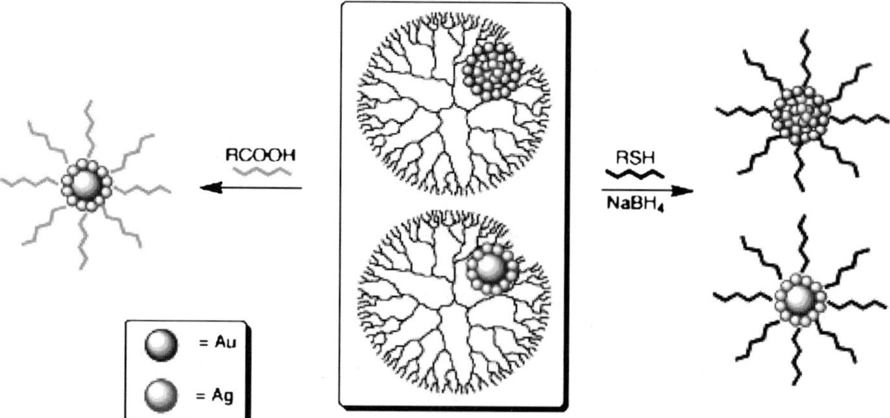

Scheme 4 Selective extraction of bimetallic AuAg nanoparticles

ries of AuAg core–Au shell DENs [G6 – OH[$Au_{27.5}Ag_{27.5}$](Au_n), where n = 95, 254, 450], the nanoparticles can be extracted from aqueous DENs into hexane solution using n-dodecanethiol. However, no extraction occurred when n-undecanoic acid was used as the extraction/protection agent, confirming Au dominates the nanoparticle shell.

This study shows that it was possible to selectively extract the nanoparticles from the dendrimer templates based on the metals present on the

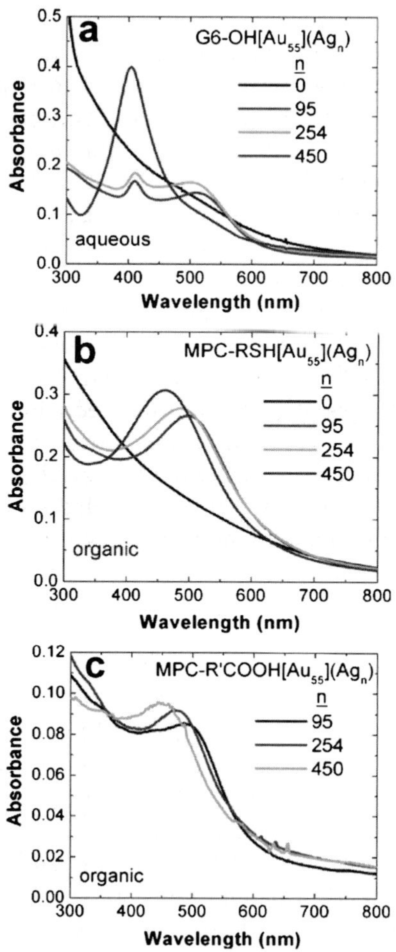

Fig. 4 UV-vis absorbance spectra of G6 – OH(Au_{55}) seeds and the G6 – OH[Au_{55}](Ag_n) (n = 95, 254, 450) series of core/shell bimetallic nanoparticles **a** before extraction, **b** after extraction with n-dodecanethiol in hexane, and **c** after extraction with n-undecanoic acid in hexane. The dendrimer concentration was 2.0 μM for all the starting DEN solutions. Reprinted with permission from J Am Chem Soc, 2005, 127, 1015–1024. Copyright 2005 American Chemical Society

particle exterior. That is, core/shell nanoparticles having Ag shells can be extracted with *n*-dodecanethiol in the presence of a reducing agent and with *n*-undecanoic acid in the absence of a reducing agent. DENs with Au shells, which can be extracted with *n*-dodecanethiol in the presence of a reducing agent, cannot be extracted with *n*-undecanoic acid under any circumstances. Therefore, although the experiment was not reported, it is likely possible to selectively extract mixtures of bimetallic nanoparticles based on surface metal compositions. This selective extraction technique is a potentially convenient tool for characterizing the composition of bimetallic DEN surfaces.

A few aspects of this study deserve comment. In the cases where well-mixed $G6 - OH(Au_n Ag_{55-n})$ ($n = 14, 27.5, 41$) DENs were used in the extraction with *n*-undecanoic acid, 25% of the atoms in each particle had to be Ag, otherwise the extraction did not proceed. Presumably, the nanoparticles need to have sufficient Ag content in order to interact with the *n*-undecanoic acid. This result is interesting in the sense that it could prove useful in future research discriminating between well-mixed and core/shell bimetallic DENs.

In the case of the $MPC - RSH[Au_{55}](Ag_n)$ systems, it appears that there may be some changes to the nanoparticles upon extraction. Evidence for this appears in the electronic spectra of the nanoparticles (Fig. 4), as the plasmon bands before and after extraction were somewhat different. The spectral differences coupled with the known miscibility of Au and Ag [41] and the strength of the Au–thiol interaction in SAMS [42], suggest that some degree of scrambling of the two metals may occur during the extraction. Evidence for similar surface sub-surface metal exchange is also found in heterogenized systems (Sect. 7). Consequently, the final morphology of the MPC cannot be unequivocally assigned to a core-shell structure.

5
Homogeneous Catalysis

Homogeneous catalysis is, of course, a major field in it's own right, as catalytic transformations are important synthetic tools. However, catalysis is also a potentially sensitive probe for nanoparticle properties and surface chemistry, since catalytic reactions are ultimately carried out on the particle surface. In the case of bimetallic DENs, catalytic test reactions have provided clear evidence for the modification of one metal by another. DENs also provide the opportunity to undertake rational control experiments not previously possible to evaluate changes in catalytic activity as a function of particle composition.

As Table 1 shows, several bimetallic DENs have been employed as homogeneous catalysts, predominately for hydrogenations. The most detailed studies have been with allyl alcohol hydrogenation and the partial hydrogenation of 1,3-cyclooctadiene (1,3-COD) test reactions. In these studies, turnover frequencies (TOFs), which can be normalized per mole of nanoparticles, can be compared as a function of the metallic atomic ratio in the DENs. Comparison to physical mixtures of monometallic DENs with the same net atomic ratio allows investigators to directly compare both the magnitude and direction of changes to rationally prepared control materials.

Figure 5 shows a plot of TOFs for the partial hydrogenation of 1,3-COD by PdRh DENs compared to TOFs for physical mixtures of Pd and Rh monometallic DENs as a function of mol % Rh. As the mol % of Rh in the bimetallic DENs was increased, an increase in the TOF was observed that was *greater* than that of the physical mixtures. Importantly, the average particle size and distribution did not change as the mol % Rh increased, which was used to rule out the possibility that the TOF enhancement was a consequence of a systematic decrease in particle size. This allows for the conclusion that the bimetallic DENs are truly intimately mixed bimetallic nanoparticles and that a "synergistic" effect is responsible for the catalytic rate enhancement.

Most of the homogeneous catalysis studies have reported some degree of catalytic rate enhancement when metals are intimately mixed in bimetallic nanoparticles. This synergistic effect was observed in the hydrogenation of allyl alcohol by PdAu [21] and PdPt [19] DENs, as well as the reduction of p-nitrophenol by AuAg [30] DENs. One particularly noteworthy study of this synergistic effect compared allyl alcohol hydrogenation by $G6 - Q_{116}(Pd_{55+n})$ to $G6 - Q_{116}[Au_{55}](Pd_n)$ for values of n from 0 to 455 (Fig. 6). This study examines the effect of particle size and morphology on catalytic activity, and highlights the type of structural study for which DENs are uniquely suited.

The first interesting observation from this study is that TOF, normalized for the total number of metal atoms, actually increased with n. As particle size increases, the fraction of surface atoms decreases, so faster reaction rates are

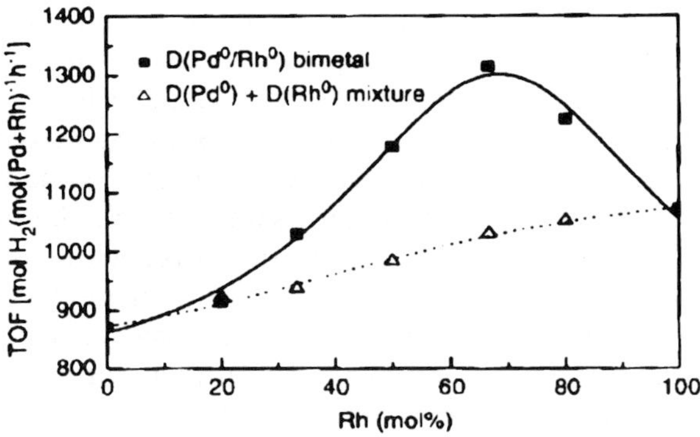

Fig. 5 Dependence of the catalytic activity of the dendrimer-encapsulated PdRh bimetallic nanoparticles on its composition in partial hydrogenation of 1,3-cyclooctadiene. Reprinted with permission from J Mol Catal, A 2003, 206, 291–298. Copyright 2003 Elsevier

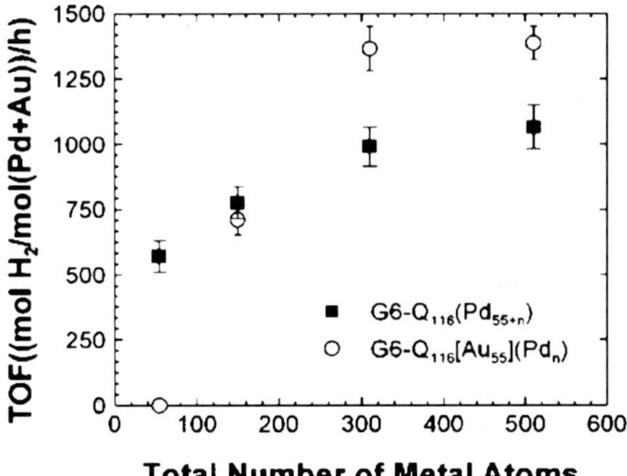

Fig. 6 Turnover frequencies (TOFs) for the hydrogenation of allyl alcohol using $G6 - Q_{116}(Pd_{55+n})$ and $G6 - Q_{116}[Au_{55}](Pd_n)$, which was prepared using the sequential-loading method, for $n = 0$, 95, 255, 455. Conditions: 22 °C, substrate : metal = 3300 : 1, [Pd + Au] = 150 μM. The ■ represent TOF data for Pd-only DECs, while the ○ represent data for the bimetallic DECs. Reprinted with permission from J Am Chem Soc, 2004, 126, 15583–15591. Copyright 2004 American Chemical Society

actually catalyzed by fewer surface atoms. It is tempting to conclude that the smallest nanoparticles are inherently less active for alkene hydrogenation in solution. However, the presence of the dendrimer makes it difficult to draw this conclusion. The dendrimer clearly plays an important role in mass transfer to the nanoparticle catalyst (this can be used advantageously to selectively hydrogenate substrates with different steric properties) [16]. Additionally, as the extraction experiments show, bonding between the metal surface and dendrimer interior amine & amide groups is important for nanoparticle stabilization. Consequently, smaller nanoparticles, which have a higher fraction of surface atoms but fewer total surface sites, could simply be more fully passivated by the dendrimer. In other words, as particle size increases, the number of free surface atoms may increase, making for a more active catalyst.

Figure 6 also shows that the Au core nanoparticles are more active catalysts than pure Pd particles containing the same total number of atoms. Since pure Au nanoparticles are inactive for alkene hydrogenation, it is difficult to attribute this enhancement to anything other than a synergistic modification of Pd by Au. The nature of the synergistic rate enhancements, which have been well documented in the homogeneous and heterogeneous catalysis literature, are of fundamental interest and importance [43]. Particle size affects both surface geometries (e.g. curvature, size of extended planar surface) and electronic structure [44]. A dopant metal can potentially affect both of these parameters by donating or withdrawing electron density or epitaxially templating altered surface arrangements. Homogeneous catalysis studies have highlighted the magnitude and direction of these effects (at least for aqueous alkene hydrogenation), but assessing the relative importance of structural and electronic effects in the presence of the PAMAM dendrimer is difficult in these systems.

6
Dendrimer Deposition and Thermolysis

The PAMAM dendrimer template offers a variety of advantages for catalysis. The ability to functionalize dendrimers makes it possible to utilize them in a variety of polar, non-polar, and fluorous solvents, and even supercritical CO_2 [16]. Because the dendrimer acts as a porous membrane, DENs can also be used to selectively hydrogenate terminal vs. internal alkenes [16]. These properties also present challenges for evaluating nanoparticle properties and for interpreting catalysis results for bimetallic catalysts. Nanoparticle surface geometric and electronic properties are extremely difficult to probe in solution, particularly when the dendrimer inhibits access by various probe molecules. Further, the number of "bonds" between nanoparticle surfaces and dendrimer amine and amide groups is essentially unknown. In cases where the dendrimer may preferentially bind one metal over another, stoichiometries and activities are difficult to evaluate, thus making it extremely difficult to interpret catalysis results in terms of particle composition.

Evaluating dendrimer templated nanoparticles in the absence of the dendrimer provides opportunities for insights into these new materials. In order to pursue these investigations, it is first necessary to immobilize DENs onto an appropriate substrate and to gently remove the dendrimer shell see Scheme 5. Opportunities for controlling nanoparticle size and composition make DENs potentially important precursors for heterogeneous catalysts and electrocatalysts, and DEN deposition and thermolysis are similarly critically important steps in pursuing these applications [45].

Wetness impregnation methods can be used to deposit DENs onto a variety of porous oxide supports, although this often requires concentrating DENs solutions to the point where dendrimer agglomeration may become problematic. Depending on the desired substrate [inorganic oxides [23, 24, 46], electroactive carbons [47], planar Au [48]] and the dendrimer terminal group, a variety of chemical deposition options are also available. Bimetallic DENs prepared using hydroxyl-terminated PAMAM dendrimers can be deposited by a variety of "slow adsorption" techniques, in which DENs are stirred with an oxide support at appropriate pH for approximately 24 hours [23, 24, 46]. This adsorption process seems to be controlled by the oxide nanoparticle interactions. Alternately, sol-gel chemistry has also been used to immobilize hydroxyl-terminated mono- and bimetallic DENs [49–51].

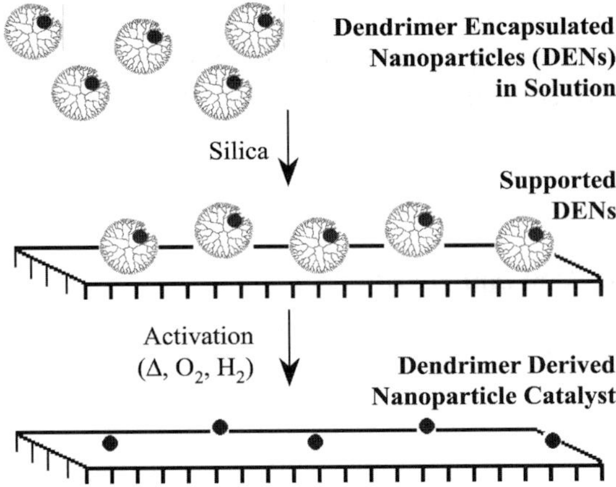

Scheme 5 Heterogeneous catalyst preparation from DENs

In the absence of a solvent, supported, intact DENs are completely inactive catalysts and do not bind CO. Presumably, upon drying, the organic dendrimer collapses onto the nanoparticle, poisoning the metal surface [52, 53]. Consequently, the dendrimer template must be removed in order for supported DENs to be used as heterogeneous catalysts. Ideally, activation conditions should be forcing enough to remove or passivate the organic material, yet mild enough not to induce particle agglomeration. Surface particle agglomeration or sintering processes are extremely temperature dependent [54, 55], so minimizing activation temperatures is crucial for ultimately correlating supported catalyst properties with synthetic methodologies and particle properties.

Several studies have shown that the amide bonds that comprise the PAMAM dendrimer backbone are relatively unstable and begin decomposing at temperatures as low as 75 °C [45, 50, 52, 56–58]. The low onset temperature of dendrimer decomposition is not surprising given that PAMAM dendrimers can undergo retro-Michael addition reactions at temperatures above 100 °C [16]. Far more forcing conditions are required to fully activate the catalysts, which suggests that the dendrimer decomposes into various surface species that continue to poison the nanoparticle surfaces.

The specific activation conditions required for an individual catalyst likely depend on the metal and support, but 300 °C appears to be somewhat of a watershed temperature. Activation at temperatures above 300 °C generally coincides with loss of metal surface area due to sintering [52, 56, 57]. The metal loading, dendrimer loading, and metal:dendrimer ratios also impact activation conditions, suggesting that it may be necessary to optimize activation conditions for individual catalysts. In most cases, treatments at 300 °C have resulted in little to no particle agglomeration [45]. Activation temperatures as low as 150 °C can also be used when CO is added to the treatment gas [59]. In this treatment, CO serves as a protecting group for the nanoparticles by preventing dendrimer decomposition byproducts from poisoning the metal surface.

7
Activated Supported Nanoparticles

7.1
Infrared Spectroscopy

Infrared spectroscopy of adsorbed CO is a useful characterization tool for dendrimer-templated supported nanoparticles, because it directly probes particle surface features. In these experiments, which are performed in a standard infrared spectrometer using an in-situ transmission or DRIFTS cell, a sample of supported DENs is first treated to remove the organic dendrimer. Samples are often reduced under H_2 at elevated temperature, flushed with He, and cooled to room temperature. Dosing with CO followed by flushing to remove the gas phase CO allows for the spectrum of surface-bound CO to be collected and evaluated. Because adsorbed CO stretching frequencies are sensitive to surface geometric and electronic effects, it is potentially possible to evaluate the relative effects of each on nanoparticle properties.

Infrared spectroscopy of adsorbed CO has been used to investigate several dendrimer-templated PtAu and PtCu catalysts. PtAu nanoparticles from $G5 - OH(Pt_{16}Au_{16})$ prepared via Cu displacement have been prepared on a variety of oxide supports (silica, alumina, titania) [24, 46]. For all the supports, bands assigned to atop $Au - CO$ and $Pt - CO$ were observed (Fig. 7). Heating the samples under He flow caused substantial changes in the $Pt - CO$ bands as $Au - CO$ desorbed, providing conclusive evidence for the intimate mixing of the two metals in individual particles. The infrared data, coupled with TEM data and CO oxidation catalytic activity indicated that bimetallic $Pt - Au$ particles might exchange surface and subsurface atoms to maximize interactions with strongly binding substrates, such as CO.

A complementary study evaluated composition effects on dendrimer-templated PtCu nanoparticles [23]. Although $Cu - CO$ bands were not observed[1], a similar red shift in the $Pt - CO$ stretching frequency to the PtAu system was observed, indicating the presence of well-mixed bimetallic nanoparticles throughout the composition range. Infrared spectroscopy of CO adsorbed on both the PtAu and PtCu catalysts showed that the shifts in the CO stretching frequency upon Cu or Au incorporation were small relative

---

[1] $Cu(0) - CO$ bands are weak, so this result is consistent with the presence of reduced Cu on the particle surface [23].

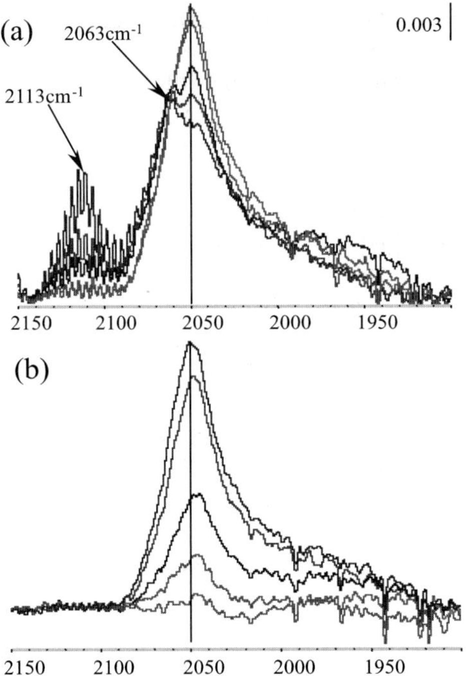

Fig. 7 Infrared spectroscopy during CO desorption from $Pt_{16}Au_{16}$ **a** 30 [*blue*], 70, 90, and 120 [*red*] °C and **b** 120 [*blue*], 150, 170, 180, and 190 [*red*] °C. Reprinted with permission from J Am Chem Soc, 2004, 126, 12949–12956. Copyright 2004 American Chemical Society

to the magnitude of dipole coupling effects [23, 24, 46, 60]. These results indicate that electronic effects (electron donation from one metal to another) are likely to be minimal for these systems [23].

7.2
Heterogeneous Catalysis

Heterogeneous catalysis also directly probes the surface properties of supported nanoparticles, and has been employed for dendrimer-templated PtAu [24, 60], PdAu [51], and PtCu [23] nanoparticles. Similar to the homogeneous catalysis studies, all three metal systems have shown synergism in catalytic activity CO oxidation catalysis, with the bimetallic catalysts being more active than any of the corresponding monometallic catalysts. For the PtCu system, the catalytic rate enhancement does not show the same maximum in activity as a function of metal content as in homogeneous hydrogenation studies [23]. This is likely due to differences in the reactions with CO oxidation active sites accounting for a small fraction of the total surface sites.

Table 2 Catalytic activity of dendrimer-templated heterogeneous PtCu catalysts

Catalyst	% Pt	Pt:Cu [a]	CO oxidation rate @ 60 °C [b]	E_{app} [c] (kJ/mol)	Toluene hydrogenation rate @ 60 °C [d]
Pt_{45}	0.19	–	180	67	1300
$Pt_{30}Cu_{15}$	0.17	1.7	350	46	460
$Pt_{23}Cu_{23}$	0.12	0.85	360	49	230
$Pt_{15}Cu_{30}$	0.057	0.43	370	47	< 5

[a] determined via AA spectroscopy
[b] mol CO/mol Pt/min $\times 10^3$
[c] Apparent activation energies from Arrhenius plots (kJ/mole)
[d] mol CO/mol Pt/min $\times 10^3$

Toluene hydrogenation catalysis has also been used to investigate dendrimer-templated PtCu nanoparticles [23]. In contrast to homogeneous hydrogenations (the PdAu system is most similar), incorporation of Cu into Pt nanoparticles had a substantial poisoning effect on toluene hydrogenation catalysis by Pt. As Table 2, shows, catalytic rates, which were normalized per total mole of Pt, drop by more than two orders of magnitude with greater Cu incorporation in the bimetallic particles. This result also contrasts with heterogeneous CO oxidation results for the same catalysts, which showed enhanced activity upon Cu incorporation.

These results were interpreted in terms of a substantial surface enrichment in Cu, driven by Cu's lower heat of sublimation [23]. The reactivity of these catalysts for CO oxidation, and the clear spectroscopic evidence for surface Pt – CO species indicate that, at least for the heterogeneous systems, particle surface stoichiometries are very sensitive to metal–adsorbate interactions. Similar arguments were presented for the PtAu/silica system, in which monometallic Au particles severely sinter under dendrimer removal conditions. In this case, the retention of small bimetallic particles after activation was attributed to the strength of Pt–silica interactions, which effectively anchored the bimetallic nanoparticles to the support [24].

8
Summary and Links Between Homogeneous and Heterogeneous Catalysis

In a sense, PAMAM dendrimers can be thought of as "nanoreactors" for preparing bimetallic nanoparticles. Several synthetic methodologies are now available for preparing a wide variety of bimetallic nanoparticles on the order of 1–3 nm. Beyond well-mixed nanoparticles, it is also possible to selectively prepare core-shell particle morphologies, with either mono- or bimetallic cores. With some shell metals and long-chain thiols or acids, it is also possible to extract nanoparticles from the dendrimer interior into organic solvents. Electron microscopy techniques, as well as UV-visible and X-ray photoelectron spectroscopies have been valuable tools in characterizing the bimetallic DENs.

The chemical and catalytic properties of the bimetallic DENs are of particular interest. Homogeneous catalysis studies generally show that bimetallic nanoparticles are more active hydrogenation catalysts than their monometallic counterparts. Synergism is also observed in heterogeneous CO oxidation catalysis. Synergism is not general, however, as heterogeneous toluene hydrogenation catalysis is severely poisoned when Cu is incorporated into Pt nanoparticles.

The source of changes in catalytic activity is of fundamental importance, and dendrimer-templated nanoparticles offer a new means of evaluating the relative influence of structural and electronic effects. Solution-phase particle size studies on alkene hydrogenation indicate anti-pathetic (activity increasing with particle size) dependence. Such homogenous studies are uncommon, but have long-standing precedent in the supported catalyst literature [61]. Heterogeneous alkene hydrogenation can be either structure sensitive (surface reaction rate depends on particle size and/or surface geometry) or structure insensitive (surface reaction rates similar over wide particle size ranges and surface geometries), depending on the alkene [62]. The presence of the dendrimer makes it difficult to unambiguously attribute the nature of the particle size effect on catalysis, but future combined homogeneous and heterogeneous catalytic studies may offer opportunities to sort this out.

Additionally, DENs offer new opportunities to investigate, understand, and evaluate the relative influence of structural and electronic effects in bimetallic catalysts. This fundamental understanding is critical if the goal of controllably tuning bimetallic catalyst properties is to be realized. Studies of CO adsorbed on heterogenized nanoparticles provide preliminary evidence that, at least for the systems studied thus far, electron donation from one metal to another may be small. This, in turn, indicates that rate enhancements may be due to changes in surface or particle geometries when a second metal is incorporated. There are still relatively few studies, however, and these conclusions can only be considered preliminary at this stage.

Another interesting and potentially important property of these particles is their potential to exchange surface and subsurface atoms. Extraction studies with AuAg particles suggest that some metal scrambling may occur when strongly binding thiols are introduced to the system. Similarly, CO seems to be able to draw Pt atoms to the surface of heterogenized PtAu and PtCu nanoparticles. It is unclear how general this property is, or if it is only applicable to certain metal systems, but it is clearly important to understanding nanoparticle dynamics in any potential application.

Acknowledgements The authors gratefully acknowledge the Robert A. Welch Foundation (Grant number W-1552) for financial support of this work. We also thank Prof. Dick Crooks and his research group for their valuable discussions.

References

1. Fisher M, Vogtle F (1999) Angew Chem Int Ed 38:884
2. Fréchet JMJ, Tomalia DA (eds) (2001) Dendrimers and other Dendritic Polymers. Wiley, West Sussex, UK
3. Kreiter R, Kleij AW, Gebbink RJM, van Koten G (2001) Top Curr Chem 217:163–199
4. Zeng F, Zimmerman SC (1997) Chem Rev 97:1681–1712
5. Bosman AW, Janssen HM, Meijer EW (1999) Chem Rev 99:1665–1966
6. Crooks RM, Zhao M, Sun L, Chechik V, Yeung LK (2001) Accts Chem Res 34:181–190
7. Twyman LJ, King ASH, Martin IK (2002) Chem Soc Rev 31:69–82
8. Cooper AI, Londono JD, Wignall G, McClain JB, Samulski ET, Lin JS, Dobrynin A, Rubinstein M, Burke ALC, Frechet JMJ, DeSimone JM (1997) Nature 389:368–371
9. Crooks RM, Zhao M, Sun L, Chechik V, Yeung LK (2001) Accts Chem Res 34:181–190
10. Crooks RM, Lemon BI, Sun L, Yeung LK, Zhao M (2001) Top Curr Chem 212:82–135
11. Ottaviani MF, Montalti F, Turro NJ, Tomalia DA (1997) J Phys Chem B 101:158–166
12. Zhao M, Crooks RM (1998) J Am Chem Soc 120:4877–4878
13. Balogh L, Tomalia DA (1998) J Am Chem Soc 120:7355–7356
14. Zhao M, Crooks RM (1999) Adv Mater 11:217–220
15. Zhao M, Sun L, Crooks RM (1998) J Am Chem Soc 120:4877–4878
16. Scott RWJ, Wilson OM, Crooks RM (2005) J Phys Chem B 109:692–704
17. Oh S-K, Kim Y-G, Ye H, Crooks RM (2003) Langmuir 19:10420–10425
18. Pellechia PJ, Gao J, Gu Y, Ploehn HJ, Murphy CJ (2003) Inorg Chem 43:1421–1428
19. Scott RWJ, Datye AK, Crooks RM (2003) J Am Chem Soc 125:3708–3709
20. Chung Y-M, Rhee H-K (2003) Catal Lett 85:159–164
21. Scott RWJ, Wilson OM, Oh S-K, Kenik EA, Crooks RM (2004) J Am Chem Soc 126:15583–15591
22. Chung Y-M, Rhee H-K (2003) J Mol Catal A-Chem 206:291–298
23. Hoover N, Auten B, Chandler BD (2006) J Phys Chem B 110:8606–8612
24. Lang H, Maldonado S, Stevenson KJ, Chandler BD (2004) J Am Chem Soc 126:12949–12956
25. Hills CW, Mack NH, Nuzzo RG (2003) J Phys Chem B 107:2626–2636
26. Luo J, Maye MM, Petkov V, Kariuki NN, Wang L, Njoki P, Mott D, Lin Y, Zhong C-J (2005) Chem Mater 17:3086–3091
27. Scott RWJ, Wilson OM, Oh S-K, Kenik EA, Crooks RM (2004) J Am Chem Soc 126:15583–15591
28. Chung YM, Rhee HK (2004) J Colloid Interf Sci 271:131–135
29. Chung Y-M, Rhee H-K (2004) Catal Surv Asia 8:211–223
30. Endo T, Yoshimura T, Esumi K (2005) J Colloid Interf Sci 286:602–609
31. Wilson OM, Scott RWJ, Garcia-Martinez JC, Crooks RM (2005) J Am Chem Soc 127:1015–1024
32. Kim Y-G, Garcia-Martinez Joaquin C, Crooks Richard M (2005) Langmuir 21:5485–5491
33. Kreibig U, Vollmer M (1995) Optical Properties of Metal Clusters, vol 25. Springer, Berlin Heidelberg New York
34. Creighton JA, Eadon DG (1991) J Chem Soc, Faraday Trans 87:3881–3891

35. Mulvaney P (1996) Langmuir 12:788–800
36. Wagner CD, Riggs WM (1979) Handbook of X-Ray Photoelectron Spectroscopy. Perkin-Elmer Co, Minnesota
37. Garcia-Martinez JC, Crooks RM (2004) J Am Chem Soc 126:16170–16178
38. Wilson OM, Scott RWJ, Garcia-Martinez JC, Crooks RM (2004) Chem Mater 16:4202–4204
39. Tao YT (1993) J Am Chem Soc 115:4350–4358
40. Schlotter NE, Porter MD, Bright TB, Allara DL (1986) Chem Phys Lett 132:93–98
41. Shibata T, Bunker Bruce A, Zhang Z, Meisel D, Vardeman Charles F, 2nd, Gezelter JD (2002) J Am Chem Soc 124:11989–11996
42. Laibinis PE, Hickman JJ, Wrighton MS, Whitesides GM (1989) Science (Washington DC, United States) 245:845–847
43. Sinfelt JH (1983) Bimetallic Catalysts: Discoveries, Concepts, and Applications. Wiley, New York
44. Schloegl R, Abd Hamid SB (2004) Angew Chemie Int Ed 43:1628–1637
45. Lang H, Chandler BD (2005) In: Nanotechnology in Catalysis. Vol 3 (in press)
46. Lang H, Auten B, Chandler BD (2005) J Catal (submitted)
47. Ye H, Crooks RM (2005) J Am Chem Soc 127:4930–4934
48. Ye H, Scott RWJ, Crooks RM (2004) Langmuir 20:2915–2920
49. Beakley L, Yost S, Cheng R, Chandler BD (2005) Appl Catal A: General 292:124–129
50. Scott RWJ, Wilson OM, Crooks RM (2004) Chem Mater 16:5682–5688
51. Scott RWJ, Sivadinarayana C, Wilson OM, Yan Z, Goodman DW, Crooks RM (2005) J Am Chem Soc 127:1380–1381
52. Lang H, May RA, Iversen BL, Chandler BD (2003) J Am Chem Soc 125:14832–14836
53. Liu DX, Gao JX, Murphy CJ, Williams CT (2004) J Phys Chem B 108:12911–12916
54. Forzatti P, Lietti L (1999) Catal Today 52:165–181
55. Bartholomew CH (2001) Appl Catal A General 212:17–60
56. Lang H, May RA, Iversen BL, Chandler BD (2005) In: Sowa JR (ed) Catalysis of Organic Reactions. Taylor & Francis Group/CRC Press, Boca Raton, FL, pp 243–250
57. Deutsch SD, Lafaye G, Liu D, Chandler BD, Williams CT, Amiridis MD (2004) Catal Lett 97:139–143
58. Lafaye G, Williams CT, Amiridis MD (2004) Catal Lett 96:43–47
59. Singh A, Chandler BD (2005) Langmuir 21:10776–10782
60. Lang H, Auten B, Chandler BD (2006) Langmuir (in revision)
61. Che M, Bennett CO (1989) Advances in Catalysis 36:55–172
62. Somorjai GA, Marsh AL (2005) Philos T Roy Soc A 363:879–900

Top Organomet Chem (2006) 20: 121–148
DOI 10.1007/3418_034
© Springer-Verlag Berlin Heidelberg 2006
Published online: 5 July 2006

Metallodendritic Exo-Receptors for the Redox Recognition of Oxo-Anions and Halides

Didier Astruc (✉) · Marie-Christine Daniel · Jaime Ruiz

Nanosciences and Catalysis Group, LCOO, UMR CNRS No. 5802, Université Bordeaux I,
351 Cours de la Libération, 33405 Talence Cedex, France
d.astruc@lcoo.u-bordeaux1.fr

Abstract Metallodendrimers terminated by redox-active transition-metal sandwiches or iron cluster groups are efficient redox exo-receptors for the recognition and sensing of various anions including oxo-anions and halides. The large size of higher-generation dendrimers ensures optimal adsorption on surfaces, a property that is used to derivatize Pt electrodes and provide re-usable sensors subsequent to washing the salt containing the anion bound by weak supramolecular interactions.

Keywords Anion · ATP · Dendrimer · Halide · Nanoparticle · Organometallic · Redox · Sensor

1
Introduction

Sensors are mostly based on light [1, 2] or electrical [3–7] stimulus. Sensing of anions is required due to their biological implication (adenosine mono-, di- and triphosphate AMP, ADP, ATP, DNA itself, RNA etc.) and ecological problems (nitrate, phosphate, radioactive pertechnetate). The area of electrochemical recognition has been remarkably well developed by Beer through elegant studies with endo-receptors (chelates, tripods, crowns, porphyrins, calixarenes) [8–14], and the recent work by Moutet's group has also added interesting contributions, especially with ATP, a DNA fragment [15–18].

The first metallodendrimers that appeared in the early 1990s were Balzani's inorganic Ru complexes [19] and polybranched iron-sandwich complexes from our group [20–22]. Despite the redox activity of both families, however, these dendritic complexes were not adequate for sensing, because no functional group able to bind anions were attached to the redox centers. The first ferrocenyl dendrimers that were found to be suitable exo-receptors for redox recognition of anions were reported in 1996 by our group in Kahn's book dealing with *Magnetism, a Supramolecular Function* [23, 24].

The dendritic topology [25, 26] is particularly appropriate for molecular sensing and recognition for several reasons. First, the dendrimer nanoscopic size matches biological substrates suggesting optimal mutual supramolecular interaction [25–32]. The nanosize is also appropriate for surface modification by the sensor. Furthermore, dendrimers are well known for their ability to encapsulate a variety of substrates [25–29]. Finally, the comparison of dendrimers of several generations will give precious information concerning the best level of recognition, i.e. on the nature of parameters that are involved in recognition. Narrow channels whose widths are decreasing with increasing dendrimer generation are the key to successful results. Thus, dendritic effects should be observed, in other words the magnitude of recognition parameters should vary upon increasing the dendrimer generation, with the expectation of positive dendritic effects (i.e. better recognition as the generation increases). It is also hoped that selectivity will be observed, i.e. the recognition of certain anions will be optimized with a given family of dendrimers whereas changing the dendrimer structure will allow recognition of other anions [33].

2
First Electrochemical Exo-receptors.
Covalently-assembled Ferrocenyl Dendrimers:
Sensing Oxo-Anions with Positive Dendritic Effects

The first ferrocenyl dendrimers suitable for redox recognition contained amido groups attached to the ferrocenyl moiety [23, 24], so that the H-bonding interaction with oxo-anions, known from Beer's work with endo-receptors [8–14], would be efficient. At this point, it is essential to point out that mononuclear amidoferrocenes that contain a linear substituent without any special topological requirement for anion recognition display no specific effect [23, 24].

The relatively small ferrocenyl dendrimers that do not contain a functional group attached to the ferrocenyl moiety [34–57] are inefficient as well. It is the synergy between the topology (endo- or exo-receptor) and the functional group attached to the ferrocenyl group that provides substantial recognition results [23, 24]. Another key point for sensing is that amidoferrocenyl dendrimers bear ferrocenyl groups that are sufficiently remote from one another (20 σ bonds) in the dendritic structures. As a consequence the electrostatic energy involved in separating the redox potentials of all redox centers [58–60] within a dendritic structure should be negligible and at least not apparent in cyclic voltammetry. This means that the cyclovoltammograms (CVs) of such metallodendrimers show a single oxidation wave at apparently the same potential, which considerably simplifies the observation of electrochemical sensing.

The first examples of ferrocenyl dendrimers, which were successfully employed for the recognition of anions, were 9- and 18-amidoferrocenyldendrimers that were formed by reaction of chlorocarbonylferrocene with polyamine dendrimers [88]. Titrations of these metallodendrimers were carried out using n-Bu$_4$N$^+$ salts of H$_2$PO$_4^-$, HSO$_4^-$, Cl$^-$, and NO$_3^-$ and monitored by CV in CH$_2$Cl$_2$ (Fig. 1) and ^{1}H NMR. In CV, two types of behavior were recorded: (i) the appearance of a new CV wave at less positive potentials for H$_2$PO$_4^-$ while the intensity of the initial CV wave decreases (for one equiv. anion, ΔE was 220 mV with 9-Fc and 315 mV with 18-Fc), and (ii) the progressive anodic shift of the initial wave for the HSO$_4^-$, Cl$^-$ and NO$_3^-$ (for one equiv. anion, $\Delta E°$ was resp. 65 mV, 20 mV and negligible with 9-Fc and 130 mV, 45 mV and 30 mV with 18-Fc).

The dichotomy of CV behavior between strongly and weakly interacting anions had been rationalized in the seminal article by Echegoyen's and Kaifer's groups with the square Scheme 1 [61]. When the strength of the interaction between the anion and the reduced redox form (here ferrocenyl) is significant, a new wave appears, and the variation of ferrocenyl potential between the free and bound forms of Scheme 1 is related to the ratio of apparent association constants: $E°_{\text{free}} - E°_{\text{bound}} = \Delta E°(V) = 0.059 \log (K_+/K_0)$ at 25 °C. $E°_{\text{bound}}$ corresponds to the addition of one equiv. anion per ferrocenyl branch or the stoichiometric amount determined from the break points, for instance in Fig. 1.

Although only the ratio of apparent association constants is accessible in this way, measurement of K_0 by ^{1}H NMR (using the shifting NH signal) can lead to K_+ as well. For instance, with 9-Fc $K_+ = (2.2 \pm 0.2) \times 10^5$ in CH$_2$Cl$_2$. When the interaction between the ferrocenyl dendrimer and the anion is negligible, only a CV wave shift is observed. The value of K_+, the apparent association constant between the oxidized (ferrocenium) form of the dendrimer and the anion, is then directly accessible from the concentration c using the equation: $\Delta E°(V) = 0.059 \log cK_+$ at 25 °C, which gives $K_+ = 544 \pm 50$, 8500 ± 500 and $61\,000 \pm 3000$ for resp. 1-Fc, 9-Fc and 18-Fc with HSO$_4^-$ in CH$_2$Cl$_2$.

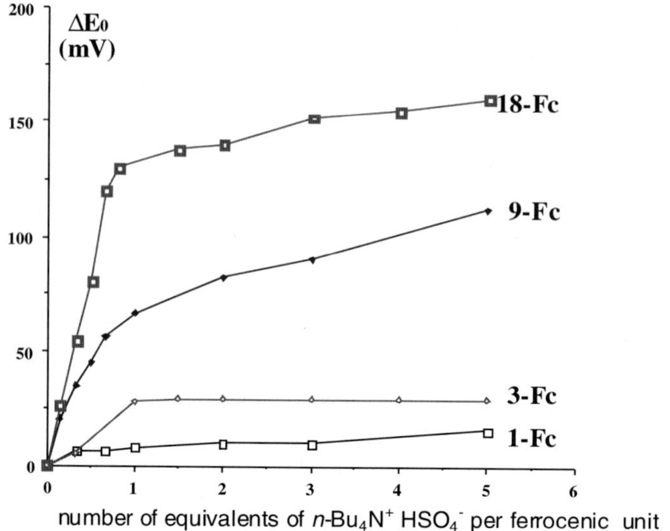

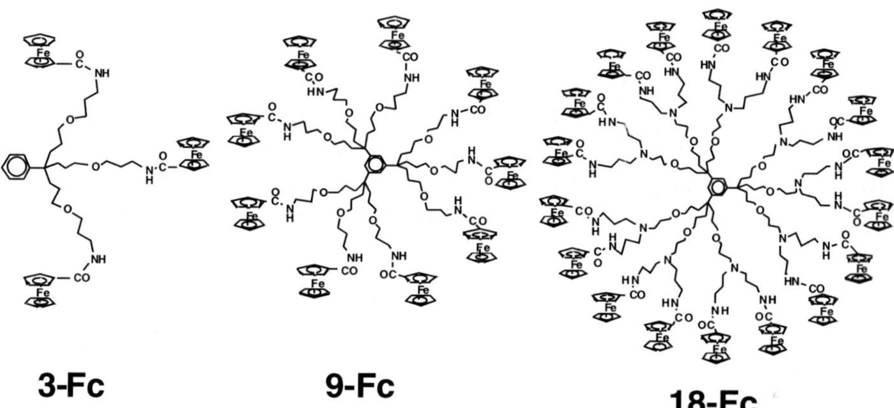

Fig. 1 Variation (shift, cf. Scheme 1, *bottom*) $\Delta E°$ of the redox potential of the ferrocenyl system recorded by CV along the titration of [n-Bu$_4$N][HSO$_4$] by mono- (1-Fc), tri- (3-Fc), nona- (9-Fc) and octadeca-amidoferrocenyl (18-Fc) compounds showing a marked dendritic effect (1-Fc is the monoamidoferrocenyl derivative [FeCp(η^5-C$_5$H$_4$CONHCH$_2$CH$_2$OPh)]) [88, 89]

The equivalence point was found by CV to correspond to the interaction of one ferrocenyl branch per equiv. anion in this series of dendrimers. The 9- and 18-amidoferrocenyl dendrimers (resp. 9-Fc and 18-Fc) were compared to monomeric (1-Fc) and trinuclear (3-Fc) amidoferrocenyl derivatives (Fig. 2), and, as indicated by the above data, a strongly positive dendritic

if K large $\Longrightarrow$ two CV waves
(the second CV wave appears upon titration)

$$C \; + \; A^- \; \underset{}{\overset{K}{\rightleftharpoons}} \; [C, A^-]$$

$-e^- \;\bigg|\; E^\circ_C \qquad\qquad\qquad -e^- \;\bigg|\; E^\circ_A$

$$C^+ \; + \; A^- \; \underset{}{\overset{K^+}{\rightleftharpoons}} \; [C^+, A^-]$$

if K small $\Longrightarrow$ only one CV wave
(that progressively shifts
upon addition of the anion)

$$C \; + \; A^- \; \underset{}{\overset{K}{\rightleftharpoons}} \; [C, A^-]$$

$-e^- \;\bigg|\; E^\circ_C$

$$C^+ \; + \; A^- \; \underset{}{\overset{K^+}{\rightleftharpoons}} \; [C^+, A^-]$$

Scheme 1 Square scheme in the cases where (i) the host-guest interaction is strong even in the reduced redox form of the host (*top*) and (ii) this interaction is negligible (*bottom*) [101]

effect (i.e. the strength of the interaction characterized by ΔE° and K_+ increases upon going to higher dendrimer generation) was characterized by the values of ΔE° and apparent association constants for the anions: ΔE° (1-Fc) $< \Delta E^\circ$ (3-Fc) $< \Delta E^\circ$ (9-Fc) $< \Delta E^\circ$ (18-Fc). The combined hydrogen bonding and electrostatic factors (attraction between the anion and the ferrocenium cation) alone cannot explain the high ΔE° values obtained with the 9-Fc and 18-Fc dendrimers, since the values found for the mono-amidoferrocenyl derivative 1-Fc are very weak (a few tens of mV). The synergy between these factors and the topology of these exo-receptors is required to provide large ΔE° and association constants (Chart 1). This positive dendritic effect can tentatively be taken into account by the narrowing of the channels between the exo-receptor redox sites that forces a tighter hydrogen-bonding interaction with increasing dendrimer generation (as shown by molecular models). This positive dendritic effect in molecular recognition contrasts with the negative dendritic effect frequently observed in catalysis whereby the reaction kinetics are lowered by the enforced steric constraints around the catalytic

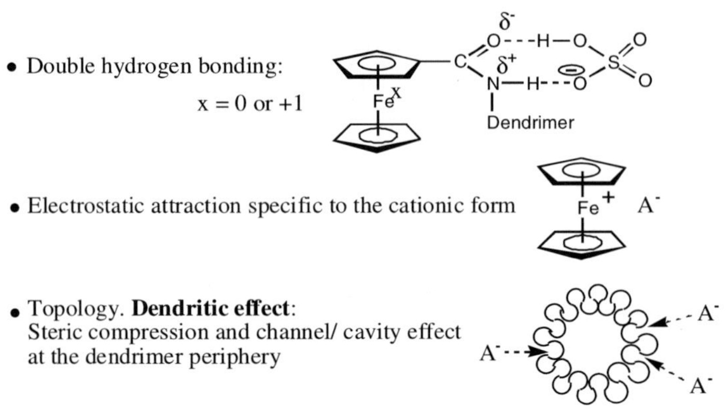

Chart 1 Factors responsible for the recognition of oxo-anions HSO_4^-, $H_2PO_4^-$ and ATP^{2-} by polyamidoferrocenyl dendrimers

metal center inhibiting its approach by substrates when the dendrimer generation increases [62, 63].

With the diaminobutane (DAB) dendrimers formulated G_n-DAB-dend-$(NH_2)_n$, amidoferrocenyl [34–42] and pentamethyl-amidoferrocenyl [64, 65] dendrimers have been synthesized (Scheme 2) for the five generations from G_1 (4 branches) to G_5 (theoretical number: 64 branches).

Recognition of HSO_4^- proceeds best in CH_2Cl_2 with the parent Cp series with, again, a positive dendritic effect (i.e. an increase of $\Delta E°$ on going to higher generations, Fig. 1) [64, 65].

In DMF, however, recognition and titration are only possible with the permethylated dendrimer series. It is subjected to another dramatic dendritic effect: the shift of the initial CV wave upon titration (weak interaction) is only observed with G_1 and a new wave (strong interaction) is observed with G_2 and G_3 [95, 96]. With amido ferrocenyl dendrimers, the oxidized ferrocenium form is not very stable due to the electron-withdrawing property of the amido group, but the electron-releasing permethylation in the Cp* ligand fully stabilizes this 17-electron form. As a result of this stabilization and the increased hydrophobicity of the Cp* dendrimers, the recognition of $H_2PO_4^-$ and ATP^{2-} is also much cleaner in the Cp* series than in the parent Cp series even if the electron-releasing character of the permethylated groups causes a slight decrease of the $\Delta E°$ values. Concerning the ATP^{2-} titration, the CVs are also clean using Cp* dendrimers, and a stoichiometry of 0.5 equiv. of ATP^{2-} per ferrocenyl branch was found, corresponding to the doubly negative phosphate charge (Fig. 3). It is noteworthy that the dendritic effect on $\Delta E°$ values in this family of dendrimers is considerably less marked and sometimes nil, a feature that was also noted with ferrocenyl-urea dendrimers of this family for the recognition of $H_2PO_4^-$ [95, 96].

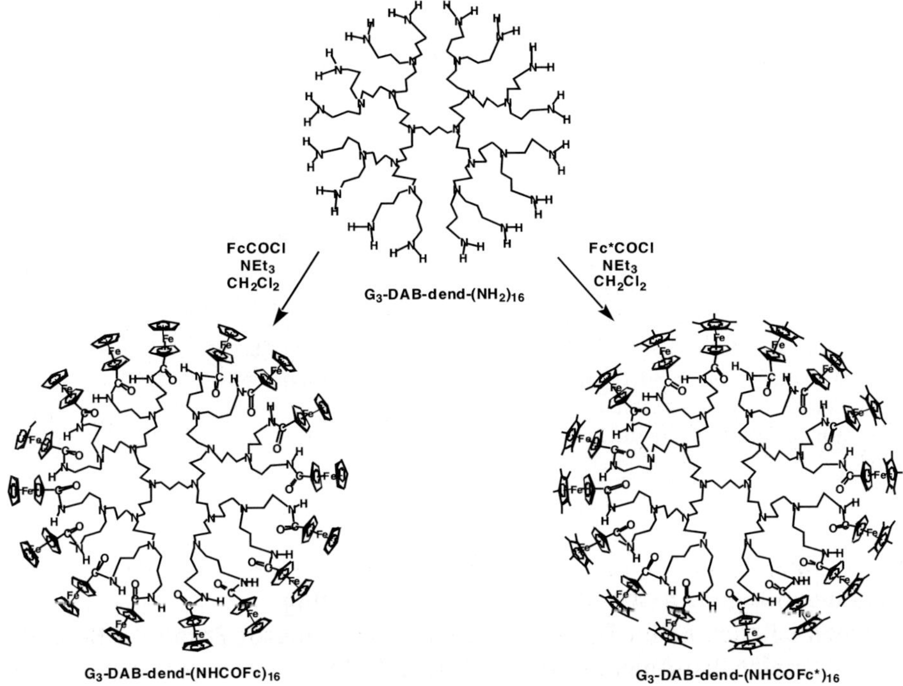

Scheme 2 Syntheses of third-generation amidoferrocenyl [34–41] and pentamethylamido-ferrocenyl dendrimers [64, 65] from the corresponding DSM polyamine dendrimer G_3-dend-$(NH_2)_{16}$. For analysis and comparison of their behavior as hosts of anions, see [64, 65]

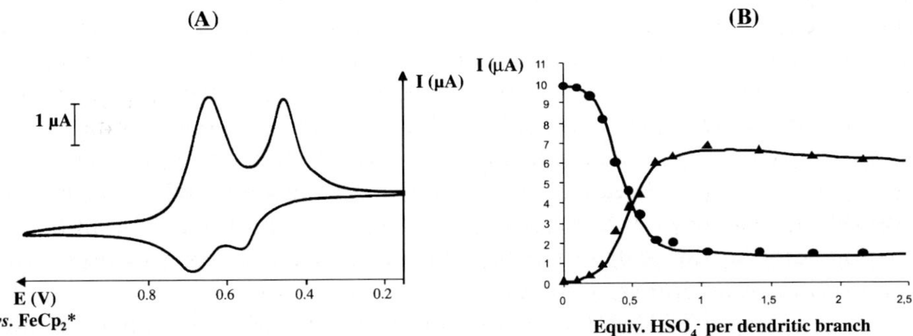

Fig. 2 Titration of a 3.3×10^{-4} M solution of the G_3-DAB-dend-$(NHCOFc)_{16}$ by a 10^{-3} M solution of [n-Bu₄N][HSO₄] in CH₂Cl₂ in the presence of 0.1 M [n-Bu₄N][PF₆], Pt anode, 20 °C. **A** Cyclovoltammogram obtained after addition of 0.5 equiv. [n-Bu₄N][HSO₄] *per* dendritic branch; **B** variation of the intensities of the initial (•) and new (▲) waves along the titration

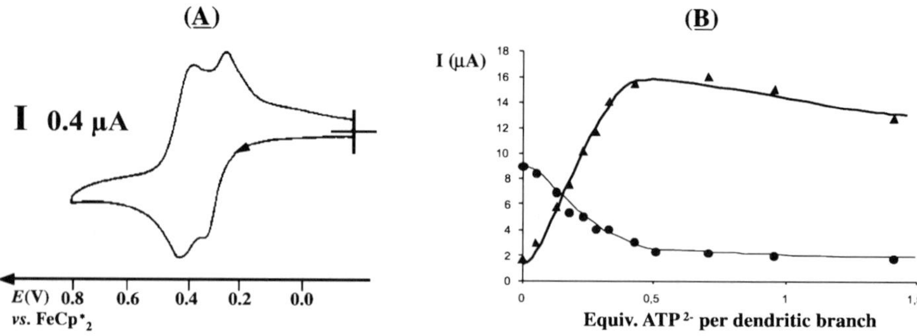

Fig. 3 Titration of a 1.25×10^{-4} M solution of G_4-DAB-dend-(NHCOFc*)$_{32}$ (Scheme 2) by a 10^{-3} M solution of [n-Bu$_4$N]$_2$[ATP] in CH_2Cl_2 in the presence of 0.1 M [n-Bu$_4$N][PF$_6$], Pt anode, 20 °C. **A** CV obtained after addition of 0.25 equiv. [n-Bu$_4$N]$_2$[ATP] per dendritic branch; **B** variation of the intensities of the initial (●) and new (▲) waves along the titration

3
Metallodendrimers Assembled by Hydrogen Bonding Between a Redox-active Dendronic Phenol and Dendritic Primary Amines also Recognize Oxo-Anions with Dendritic Effects

Supramolecular aspects of dendrimer chemistry, especially those involving hydrogen bonding have been a subject of interest in the last decade [66–71]. We were intrigued by the very simple possibility of hydrogen bonding between primary amines and phenols and by its potential use in dendrimer chemistry and molecular recognition. The tedious, time-consuming dendritic syntheses often represent an obstacle for research and use of dendrimers. Therefore the advantage of forming dendrimers with redox-active termini by simply mixing a commercial polyamine core and a redox-active dendron is obvious (Chart 2). This is especially the case if a specific exo-receptor property with positive dendritic effect can be obtained as shown above in the covalently synthesized metallodendrimers. Indeed, there were precedents for hydrogen bonding between primary amines and alcohols with tetrahedral disposition of both O and N valences and 1 : 1 stoichiometry (i.e. minimal melting point for this stoichiometry), a property that had been used in crystal engineering and chiral recognition [72–75].

Thus, mixing a DSM dendritic polyamine and a para-substituted phenol derivative leads to the replacement of the ^{1}H NMR signals of the OH protons at 5 ppm and NH$_2$ proton at 1.5 ppm by a common broad, concentration-dependent signal for these three protons located between 2.4 and 4.1 ppm. This means that supramolecular dendrimers involving reversible hydrogen bonding between the DSM polyamine and a phenol dendron form upon mixing these two components. The electrochemical time scale of the CV being

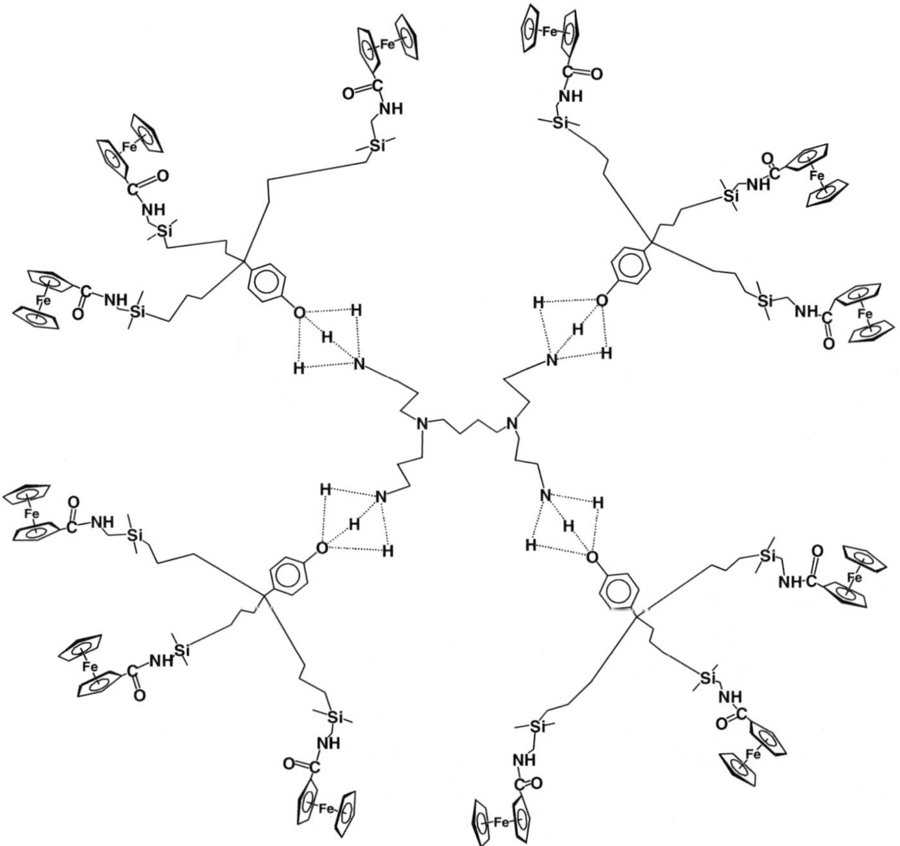

Chart 2 Arbitrary representation of the reversible hydrogen bonding between G_1-DAB-dend-$(NH_2)_4$ and a triamidoferrocenyl dendron shown by the concentration-dependent average location of the (broad) NH_2 + OH signal in ^{1}H NMR between 2.4 and 4.1 ppm vs. TMS in $CDCl_3$

much larger than that of the hydrogen-bond formation and breaking, the CV shows an average situation between the hydrogen-bonded and non-bonded dendrimer branches [76, 77].

Upon titration of $[n\text{-}Bu_4N][H_2PO_4]$ by FcCONHPr (Fc = ferrocenyl), the CV of this ferrocenyl group displays the appearance of a new, anodically shifted wave with $\Delta E° = 150$ mV; this value reaches 210 mV with the dendron $p\text{-}OH - C_6H_4C\{(CH_2)_3SiMe_2 - CH_2NHCOFc\}$ alone, and it is not changed in the presence of propylamine. It reaches 250 mV, however, with this dendron + G_1-DAB-dend-$(NH_2)_4$ and 280 mV with this dendron + G_2-DAB-dend-$(NH_2)_8$ or a higher generation polyamine dendrimer. This increase of the $\Delta E°$ value is the signature of a dendritic effect as shown previ-

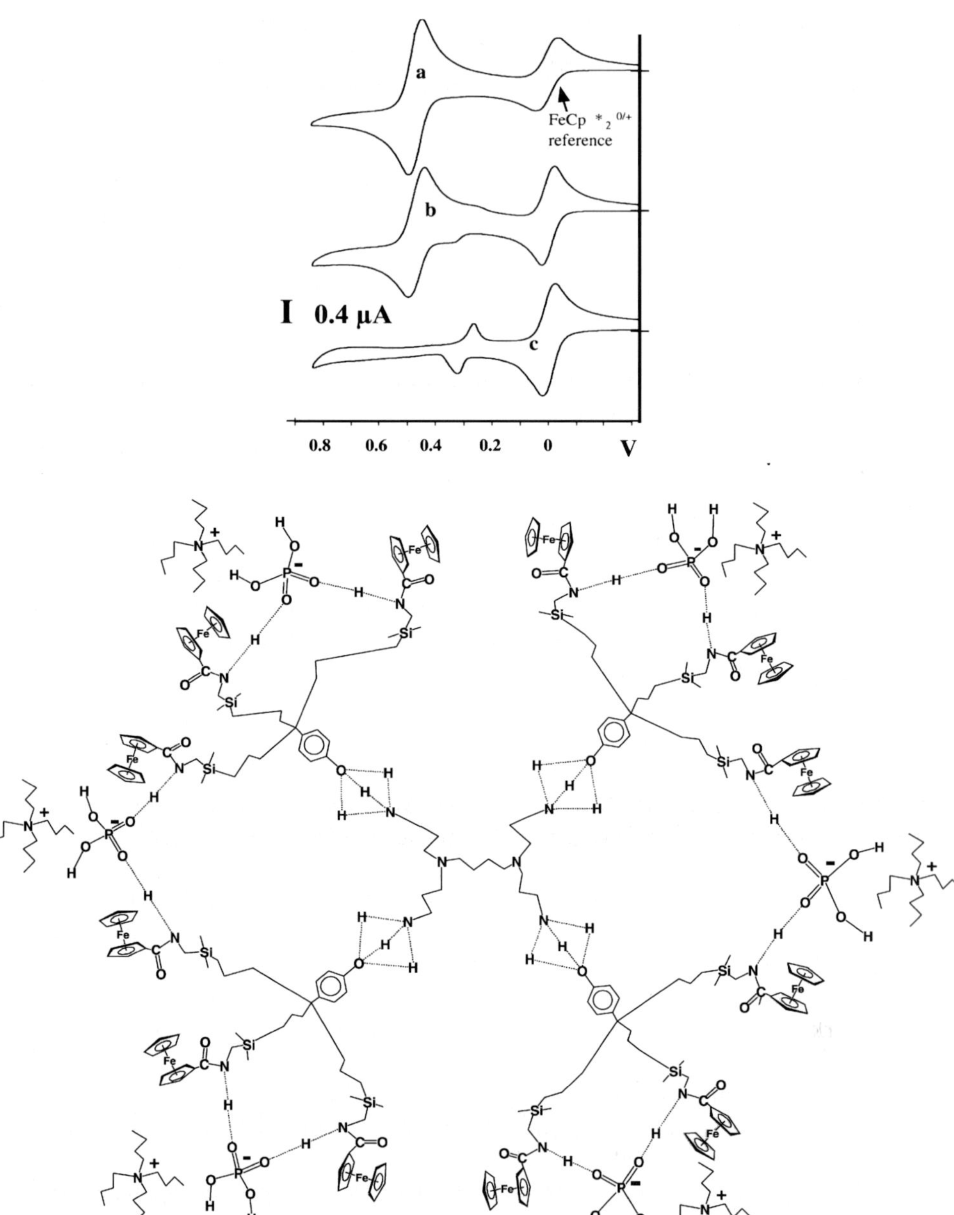

Fig. 4 *Top*: Titration of DAB-G$_1$ in CH$_2$Cl$_2$ (Pt, 0.1 M [*n*-Bu$_4$N][PF$_6$], 20 °C; reference: FeCp$_2^*$ = decamethylferrocene) by [*n*-Bu$_4$N][H$_2$PO$_4$]: **a** before addition; **b** 0.4 equiv.; **c** 0.5 equiv.; *bottom*: proposed supramolecular assembly between the triferrocenyl dendron + DAB-G$_1$ and 0.5 equiv. [*n*-Bu$_4$N][H$_2$PO$_4$] taking into account the sudden drop of CV wave intensity at the equivalence point (i.e. sudden drop of diffusion coefficient)

ously with covalent redox metallodendrimers. With the monomer FcCONHPr + n-propylamine, it is necessary to add 2.5 equiv. $[n\text{-}Bu_4N][H_2PO_4]$ to reach the equivalence point whereas one equiv. $[n\text{-}Bu_4N][H_2PO_4]$ is enough with the covalent dendrimers, because the amine strongly competes with the amidoferrocenyl group in binding $H_2PO_4^-$. With G_1, only 0.5 equiv. $[n\text{-}Bu_4N][H_2PO_4]$ is necessary, and a sudden disappearance of the initial CV wave is observed whereas the intensity of the new CV wave is much reduced. This can tentatively be accounted for by the formation of a rather stabilized dendritic supramolecular assembly in which the $H_2PO_4^-$ anion bridges two amidoferrocenyl units (Fig. 4).

The reduction of the diffusion coefficient responsible for the decrease of the CV wave intensity at the equivalence point can be explained by the sudden increase of the mass of this overall supramolecular assembly compared to the much smaller species present in solution before the equivalence point. For higher generations, the number of equiv. $[n\text{-}Bu_4N]\,[H_2PO_4]$ necessary to reach the equivalence point progressively increases again (0.8 for G_2 and 2.0 for G_3 and G_4), probably due to the fact that the steric congestion around the dendrimer partly destabilizes, for high generations, the hydrogen bonds optimized in G_1 and G_2; this is also consistent with the limit of the increase of ΔE°. This strategy has also been successfully applied to the recognition of $[n\text{-}Bu_4N]_2[ATP]$ for G_1 [77].

4
Gold nanoparticle–alkanethiolate-ferrocenyl Stars and Dendrimers are Excellent Oxo-Anion Sensors that can Provide Derivatized Electrodes

Gold nanoparticles have been known and used for centuries. Their interest lies in their biocompatibility, ease of fabrication and their manifold applications in nanotechnology including optics, electronics and even catalysis [78]. With alkylthiolate ligands, they are especially known from Mulvaney's seminal work [79, 80] and the popular and practical Brust–Schiffrin biphasic method of synthesis [81, 82]. They are very robust, quite monodisperse and easy to characterize by combining transmission electron microscopy (TEM), elemental analysis (Au/S ratio) and, as simple molecules, by standard spectroscopic methods (NMR, infrared, UV-Vis) [83]. Thus, a variety of functional alkylthiolate ligands including ferrocenylalkylthiolates have been introduced either by direct synthesis or by ligand substitution of non-functional alkylthiolates with good control of the degree of substitution [78, 84–86]. Variable proportions (7 to 38%) of amidoferrocenylundecanethiol (AFAT), a ligand deposited as self-assembled monolayers on gold surfaces by Creager [85, 86], were added into alkylthiolate-AuNPs. The original core size of diameter around 2 nm with a number of ligands about 100 remained unchanged in the course of these reactions [87].

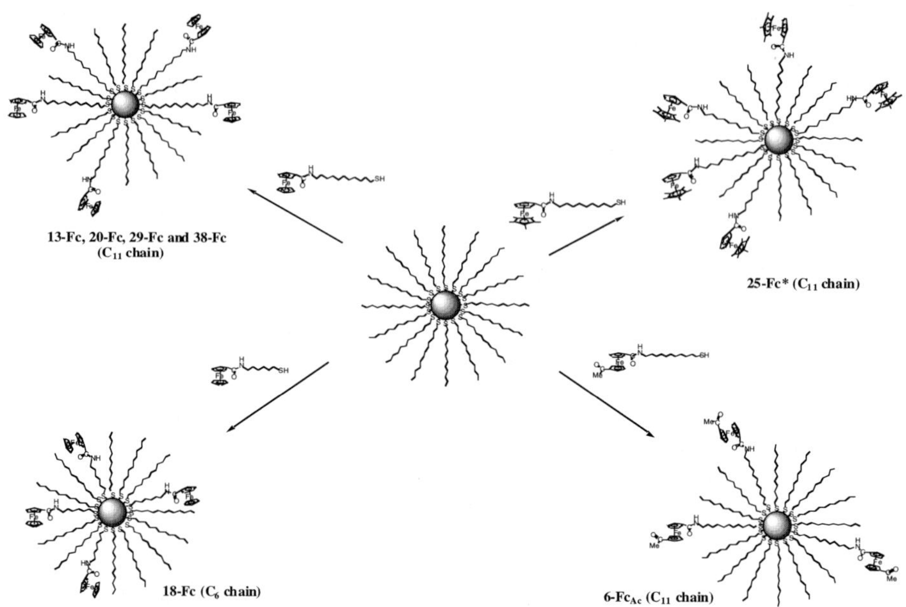

Scheme 3 Syntheses of various AFAT-AuNPs (using the ligand substitution procedure) for the selective recognition and titration of oxo-anions (for instance, 18-Fc means AuNPs with 18% AFAT ligand and 82% dodecanethiolate ligand)

As ferrocenyl-terminated dendrimers, these AFAT-AuNPs (Scheme 3) display a single CV wave in CH_2Cl_2 with a difference of potentials between the cathodic and anodic peaks that is lower than 60 mV (typically 20 mV). This indicates some adsorption due to the large size [88, 89]. Addition of $[n\text{-Bu}_4\text{N}][H_2PO_4]$ leads to the appearance of a new CV wave, and the equivalence point is reached for a 1 : 1 stoichiometry (equiv. $[n\text{-Bu}_4\text{N}][H_2PO_4]$ per AFAT branch), similar to the observations made with dendrimers. It is remarkable that the ΔE° value is as large as 220 mV and constant regardless of the proportion of AFAT ligands in the AFAT-AuNPs, corresponding to an apparent association constant ratio K_+/K_0 between $H_2PO_4^-$ and the AFAT-AuNPs of 5350 ± 550 (Fig. 5).

It is also possible to titrate $[n\text{-Bu}_4\text{N}][H_2PO_4]$ selectively in the presence of both $[n\text{-Bu}_4\text{N}][HSO_4]$ and $[n\text{-Bu}_4\text{N}]Cl$ or to titrate $[n\text{-Bu}_4\text{N}][HSO_4]$ alone, but the ΔE° value (cathodic shift) is only 40 mV with this latter anion (Fig. 6). The halides and nitrate cannot be recognized, however.

The alkyl chain length was varied without significant effect, but stereo-electronic effects of the AFAT ligand play a role: a Cp* analogue, and a heterodisubstituted 1-amido, 1′-acetyl-ferrocenyl derivative have been introduced onto AuNPs[50]. The electron-releasing Cp* ligand reduced the ΔE° value to 125 mV whereas the additional electron-withdrawing acetyl group enhances it to 275 mV (Fig. 7).

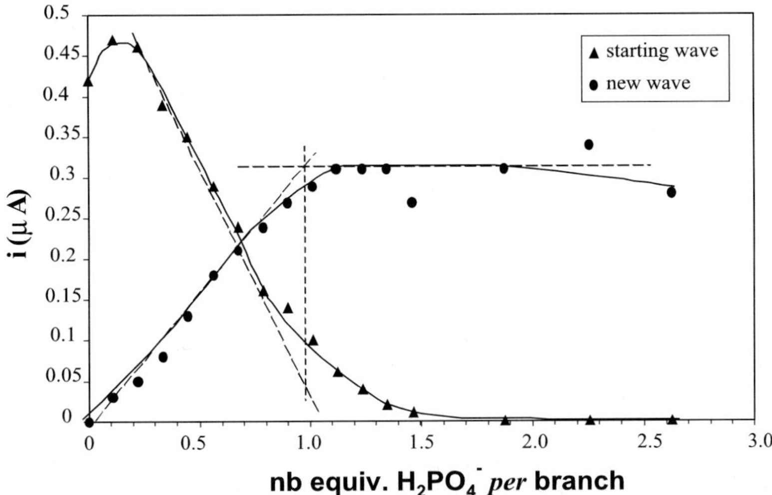

Fig. 5 Titration of [n-Bu$_4$N][H$_2$PO$_4$] with the 20-Fc AFAT-AuNPs monitored by CV (i.e. AuNPs containing 20 AFAT ligands and about 80 dodecanethiolate ligands). Decrease of the intensity of the initial CV wave (▲) and increase of the intensity of the new CV wave (•) vs. the number of equiv. of [n-Bu$_4$N][H$_2$PO$_4$] added per AFAT branch. Nanoparticles: 5×10^{-6} M in CH$_2$Cl$_2$; [n-Bu$_4$N][H$_2$PO$_4$]: 10^{-2} M in CH$_2$Cl$_2$, [n-Bu$_4$N][PF$_6$]: 0.1 M, 20 °C, reference electrode: SCE, auxiliary and working electrodes: Pt. Scan rate: 200 mV s^{-1}

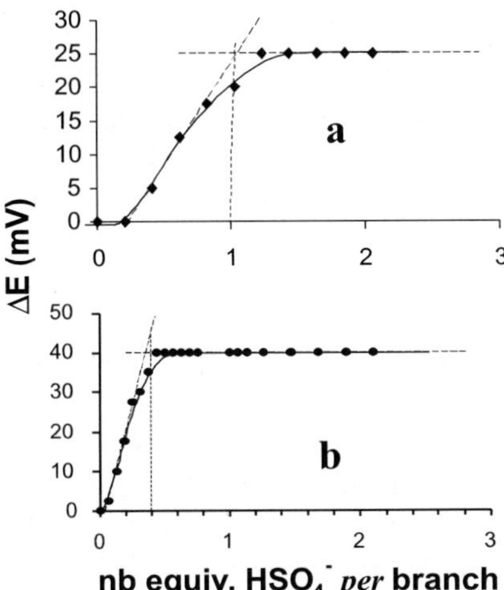

Fig. 6 Titration of [n-Bu$_4$N][HSO$_4$] by gold nanoparticles containing AFAT ligands in CH$_2$Cl$_2$. Shift of $E_{1/2}$ towards positive potentials as a function of the number of equiv. [n-Bu$_4$N][HSO$_4$] added per amidoferrocenyl branch of the colloids. **a** Titration by 13-Fc-AFAT-AuNPs: the equivalence point is 1 equiv. [n-Bu$_4$N][HSO$_4$] per amidoferrocenyl branch. **b** Titration by 38-Fc: the equivalence point is only 0.4 equiv. [n-Bu$_4$N][HSO$_4$] *per* amidoferrocenyl branch, possibly due to preferred amide–amide interactions of neighboring branches at high ligand load

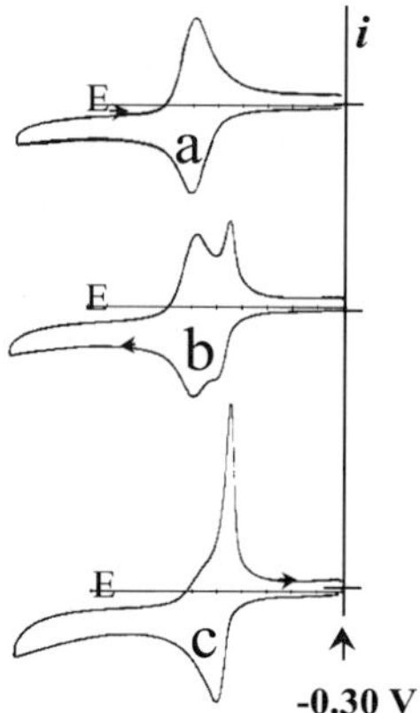

Fig. 7 Cyclic voltammograms of the 25-Fc*-AuNPs (Fc* = Cp*FeCp-). Solvent: CH_2Cl_2; reference electrode: aqueous SCE; working and counter electrode: Pt; supporting electrolyte: 0.1 M n-Bu$_4$NBF$_4$; scan rate: 200 mV s^{-1}. **a** Without [n-Bu$_4$N][H$_2$PO$_4$], **b** with 0.75 equiv. [n-Bu$_4$N][H$_2$PO$_4$] per 1-amido, 1′-pentamethyl-ferrocenyl branch; **c** with excess [n-Bu$_4$N][H$_2$PO$_4$]

Similarly, for the recognition of [n-Bu$_4$N][HSO$_4$], the presence of the Cp* ligand reduces the $\Delta E°$ value to 25 mV whereas that of the withdrawing acetyl group enhances it to 170 mV. Interestingly, the introduction of the acetyl group also transforms the CV wave shift (weak-interaction case with the parent AFAT ligand) to the formation of a new wave (strong-interaction case). In summary, the AFAT-nanoparticles and the amidoferrocenyl dendrimers show some similar trends, but detailed investigations indicate remarkable differences in the selectivity and enhanced flexibility in the design of the stereoelectronic feature around the ferrocenyl group [89]. Moreover, the AFAT-AuNPs are directly accessible whereas dendrimers require tedious multi-step syntheses. This ease of access will be further exploited with a nanoparticle-cored ferrocenyl dendrimer.

Nanoparticle-cored dendrimers are new materials that were reported for the first time in 2001 [90–92]. With ferrocenyl termini, they combine the advantages of covalent ferrocenyl dendrimers and AFAT-AuNPs. Moreover, their large size (in our case) offers the possibility to easily derivatize electrodes with the redox ferrocenyl sensors, which makes them useful for sensing [93, 94]. Thus, ferrocenyl AB$_3$ and AB$_9$ dendrons were synthesized with a monothiol for the ligand part A and three or nine amidoferrocenyl or silylferrocenyl groups for B$_3$ [90] or B$_9$ [93, 94]. The AB$_3$ dendrons were assembled into AuNP-cored dendrimers either by direct synthesis with the Brust–Schiffrin method using a mixture of dodecanethiol and tripodal thiol ligands (Scheme 4) or by the ligand-substitution method from dodecanethiol-AuNPs and a given amount of tripodal thiol ligand. The numbers of AB$_3$ dendrons introduced in this way was between five and seven in 2.3-diameter AuNPs

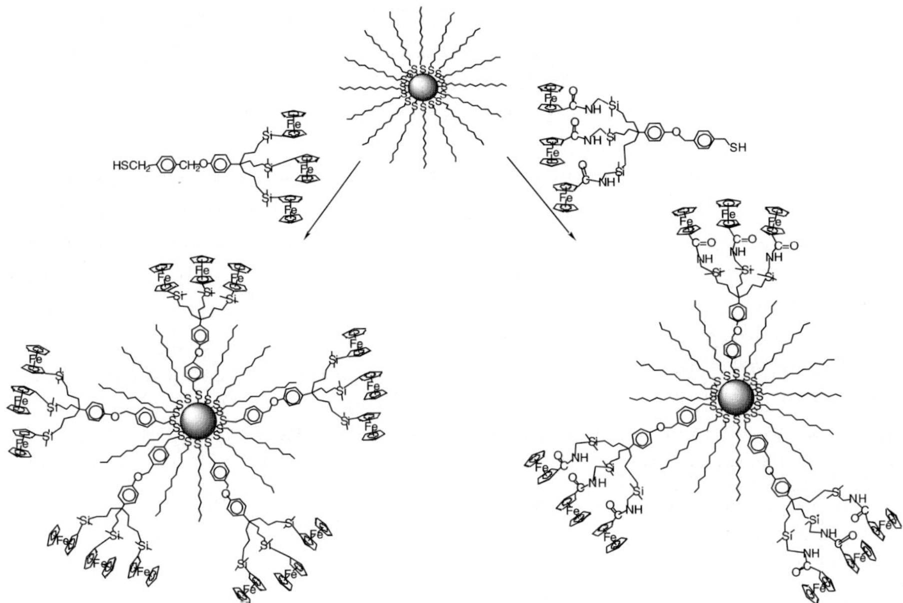

Scheme 4 General synthetic scheme for triferrocenyl-dendronized gold nanoparticles (ligand substitution method)

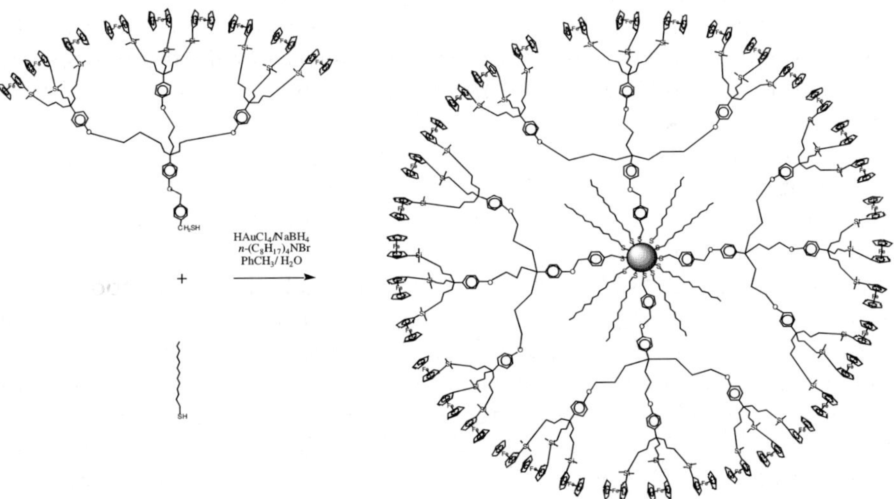

Scheme 5 Synthesis of a AuNP-cored silylferrocenyl dendrimer (direct method)

containing about 150 thiolate ligands overall. For the silylferrocenyl AB$_9$ dendrons, the ligand-substitution procedure no longer worked, presumably for steric reasons, but the direct synthesis using a mixture of dodecanethiol and

AB$_9$ thiol gave good results (Scheme 5). Two closely related AB$_9$ dendrons were used leading to an assembly of about 10% resp. 20% of AB$_9$ thiolate dendrons around 2.9-nm AuNPs bearing around 200 thiolate ligands overall. This means that these AuNPs contain about 180 resp. 360 equivalent silylferrocenyl groups at the periphery [93, 94].

The recognition experiments led to stoichiometries corresponding to one equiv. of [n-Bu$_4$N][H$_2$PO$_4$] per ferrocenyl branch with all the AuNP-cored ferrocenyl dendrimers. The $\Delta E°$ values for this anion were 200 mV with the amidoferrocenyl AB$_3$ dendron-AuNPs, 115 mV for the silylferrocenyl AB$_3$ dendron-AuNPs and 125 mV for the AB$_9$ dendron-AuNPs. Recognition studies with [n-Bu$_4$N]$_2$[ATP] gave results that were similar to those obtained with [n-Bu$_4$N][H$_2$PO$_4$] except that the stoichiometry was reached for only 0.5 equiv. [n-Bu$_4$N]$_2$[ATP] per ferrocenyl branch because of the double negative charge of ATP^{2-} (Fig. 8). The $\Delta E°$ values were usually only slightly lower for [n-Bu$_4$N]$_2$[ATP] than for [n-Bu$_4$N][H$_2$PO$_4$].

The silylferrocenyl AuNP-cored dendrimers are thus efficient sensors, because the silylferrocenium species is stable unlike the parent amidoferrocenium system. The silicon atom attached to the ferronyl group plays the role of a Lewis acid that interacts with the H$_2$PO$_4^-$ anion by hydrogen bonding. With [n-Bu$_4$N][HSO$_4$], however, the silylferrocenyl dendrimers do not provide recognition. A $\Delta E°$ value of 42 mV with $K_+ = (18 \pm 4) \times 10^3$ L mol^{-1} was obtained with this anion using the tris-amidoferrocenyl dendron-AuNPs with

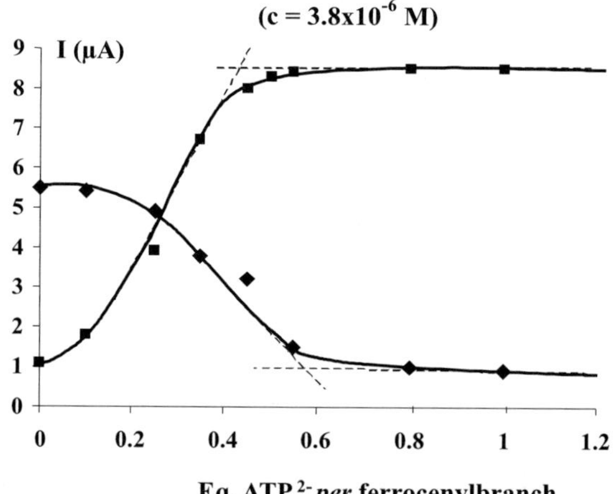

Fig. 8 Titration of ATP^{2-} with 9-Fc-thiolate dendron-AuNPs (Scheme 5). Decrease of the intensity of the initial CV wave; (♦) increase of the intensity of the new CV wave (■) vs. the number of equiv. of [n-Bu$_4$N]$_2$[ATP] added per ferrocenyl branch. Nanoparticles: 3.8×10^{-6} M in CH$_2$Cl$_2$

the same stoichiometry. The weaker interaction of the amidoferrocenyl dendrimers with $[n\text{-Bu}_4\text{N}][\text{HSO}_4]$ than with $[n\text{-Bu}_4\text{N}][\text{H}_2\text{PO}_4]$ is due to the fact that NH-oxo anion hydrogen bonding is dominant, and the negative charge on the oxygen atoms is smaller in HSO_4^- than in H_2PO_4^- (sulfur being more electro-negative than phosphorus).

Electrodes modified by ferrocenyl polymers have been known for a long time [93, 94]. More recently, Cuadrado et al. have extensively studied the derivatization of silylferrocenyl dendrimers [34–42]. Nishihara's group has reported the first example of modified electrodes with AuNPs containing ferrocenyl thiol ligands, in which the stabilization of the modified electrodes is provided by the biferrocenyl units whereas monoferrocenyl units do not afford stabilization [95, 96]. We have noticed that the larger the ferrocenyl dendrimers, the more strongly they tend to adsorb on Pt electrodes in CH_2Cl_2 solutions, and the easier it is to prepare modified Pt electrodes with ferrocenyl dendrimers by scanning around the region of the ferrocenyl potential [97]. Thus, the AuNP-cored ferrocenyl dendrimers discussed above are large and readily adsorb on Pt electrodes, forming perfectly stable derivatized electrodes upon scanning about 50 times (for saturation) around the ferrocenyl potential region. These modified Pt electrodes show remarkable changes upon introduction of a solution of $[n\text{-Bu}_4\text{N}]_2[\text{ATP}]$. The new CV wave observed has undergone a shift of potential; the anodic and cathodic peaks are no longer indentical showing that a structural rearrangement, due to changes in hydrogen bonding and electrostatic interaction, occurs in the course of the heterogeneous electron transfer. $[n\text{-Bu}_4\text{N}][\text{H}_2\text{PO}_4]$ and $[n\text{-Bu}_4\text{N}]_2[\text{ATP}]$ are selectively recognized in the presence of other anions such as HSO_4^- and Cl^-; $[n\text{-Bu}_4\text{N}][\text{HSO}_4]$ alone can also be recognized. The recognized salts can be washed from the electrode using CH_2Cl_2, but the AuN-cored ferrocenyl

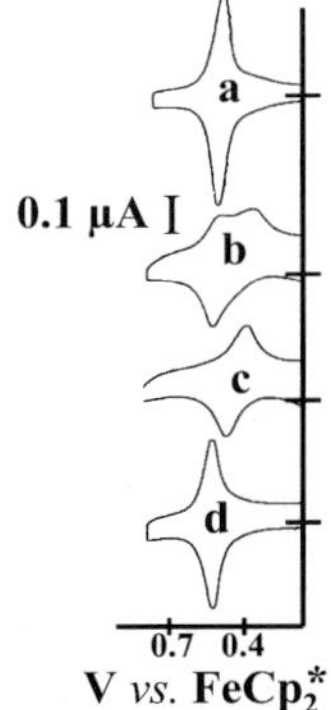

Fig. 9 Recognition of ATP^{2-} with AuNP functionalized with a nonasilylferrocenyl dendron shown in Scheme 7. CV: **a** Modified electrode alone; **b** during the course of titration; **c** with excess of $[n\text{-Bu}_4\text{N}]_2[\text{ATP}]$; **d** after removal of $[n\text{-Bu}_4\text{N}]_2[\text{ATP}]$ upon washing the modified electrode with CH_2Cl_2

dendrimers are not removed and can serve again for further analogous experiments. Such recycling can be performed several times (Fig. 9) [92].

5
Dendrimers Containing Fe$_4$ Clusters at the Periphery Recognize ATP^{2-} Better than the Model H$_2$PO$_4^-$ and also Form Derivatized Electrodes as Re-usable Sensors

The complex [Fe$_4$(μ_3-CO)$_4$Cp$_3$(η^5-C$_5$H$_4$COCl)] [98, 99], reacts with the nona-branch amino dendrimer in the presence of NEt$_3$ as shown in Scheme 6 to give the expected nona-branch amidocluster dendrimer 1 (Scheme 6). The successful functionalization of the 3rd-generation 16-NH$_2$ DAB dendrimer shown in Scheme 7 was performed using the N-succinimidyl ester of the Fe$_4$-cluster [99], giving the 16-branch amidocluster dendrimer 2 (Scheme 7) [100].

The functionalization worked smoothly with the new 27-NH$_2$ dendrimer shown in Scheme 8 [100] giving the 27-Fe$_4$ dendrimer 3. These new metallodendrimers 1, 2 and 3 are air-stable, forest-green powders. They were characterized using standard spectroscopic and elemental analyses, and AFM of 2 and 3 (Fig. 10) shows that it forms monolayers of aggregated metallodendrimers on the mica surface. Indeed, the height of 1.5 nm (2) and 2.4 nm (3) is reproducibly obtained by AFM and corresponds to the dimension of slightly flattened molecular models.

The cyclovoltammograms of the dendrimer-clusters 1, 2 and 3 in CH$_2$Cl$_2$ (Pt, 0.1 M n-Bu$_4$NPF$_6$) resemble that of the parent cluster [Fe$_4$(μ_3-CO)$_4$Cp$_4$] itself [101], the clusters being sufficiently remote from one another in the den-

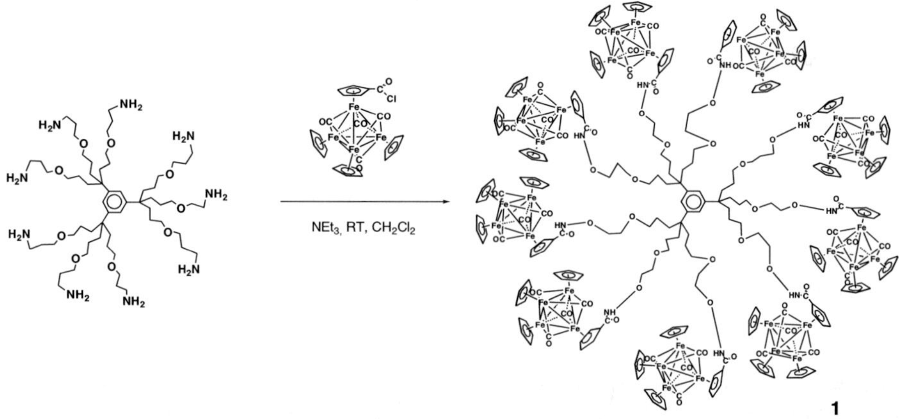

Scheme 6

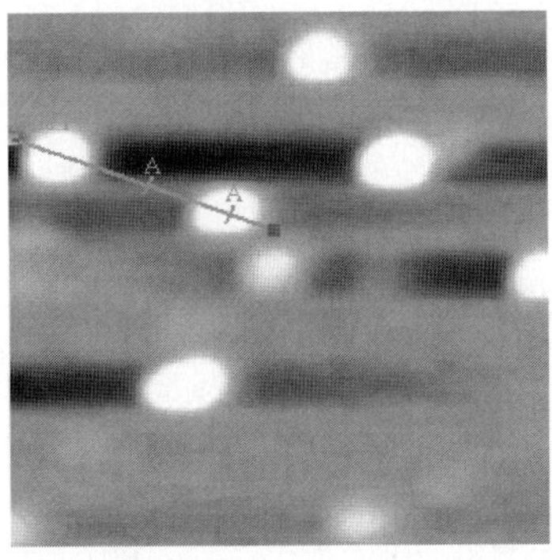
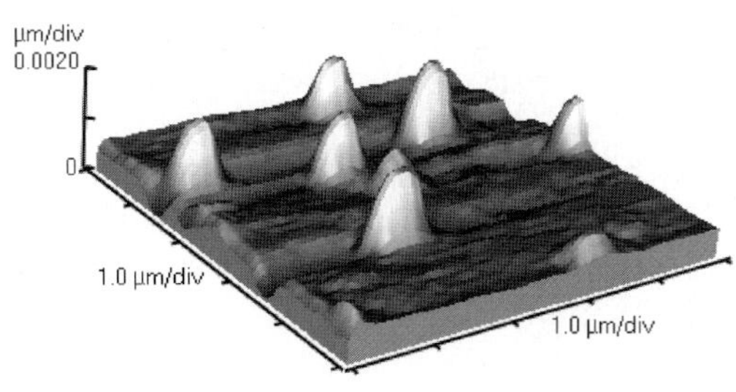

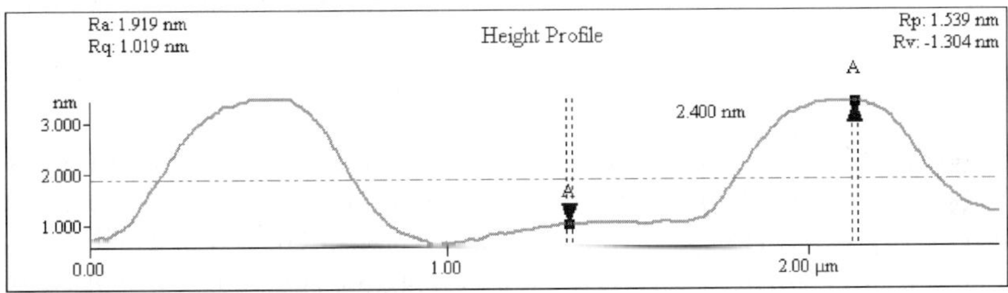

Fig. 10 AFM of the 27-Fe$_4$-cluster dendrimer **3** (see Scheme 8)

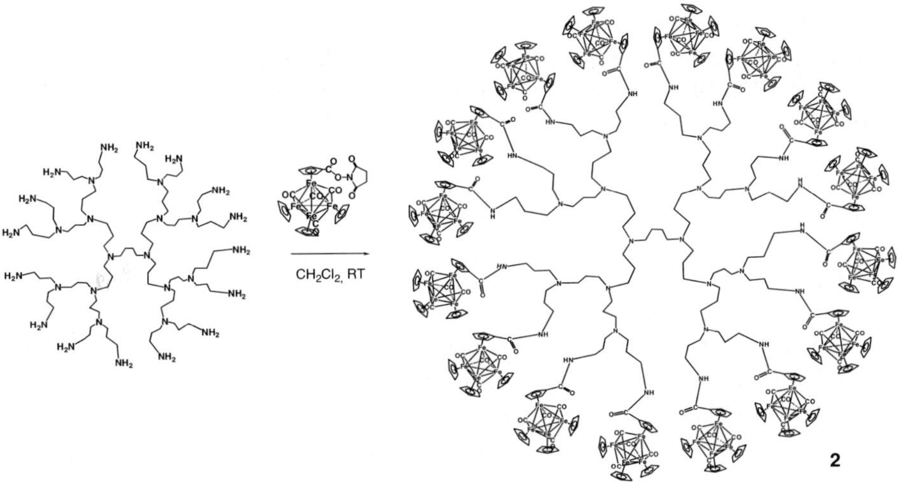

Scheme 7

drimers to render the electrostatic factor almost nil. Therefore all the redox sites corresponding to the redox change Fe$_4$ → Fe$_4{}^+$ appear in a single reversible wave (whereas the other waves Fe$_4{}^+$ → Fe$_4{}^{2+}$ and Fe$_4{}^0$ → Fe$_4{}^-$ are

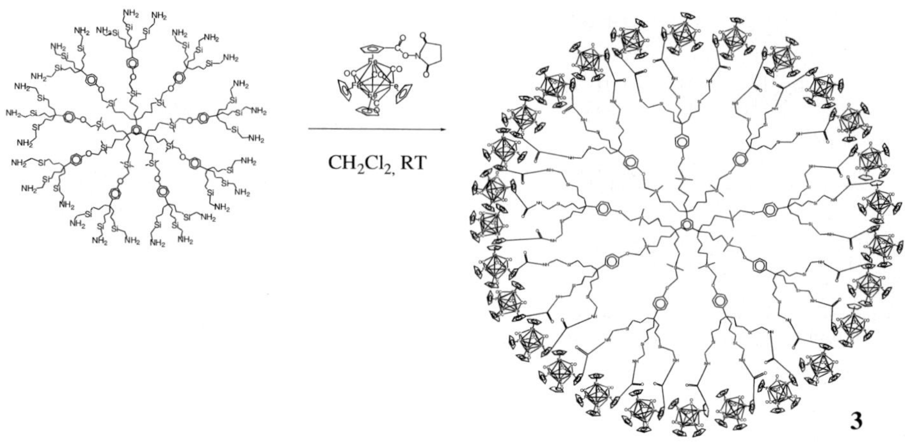

Scheme 8

not as reversible [61]. For this wave, application of the Bard–Anson equation [61] to determine the number of electrons, under conditions that avoid adsorption, provides a result of 27 ± 3 electrons in CH_2Cl_2 and DMF using $[FeCp_2^*]$ as the internal reference. This stoichiometry could be confirmed by titration of **3** in CH_2Cl_2 using 27 equiv. of green $[CpFe\eta^5\text{-}C_5H_4COCH_3][PF_6]$, which generates $[CpFe\eta^5\text{-}C_5H_4COCH_3]$ and a dark-green precipitate of $[3][PF_6]_{27}$ characterized by $\nu_{CO} = 1690\,cm^{-1}$, $55\,cm^{-1}$ higher than ν_{CO} of **3** ($1635\,cm^{-1}$). A color change of the CH_2Cl_2 solution from dark-green to red occurs at the equivalence point.

Recognition of the oxo-anions HSO_4^-, $H_2PO_4^-$ and adenosine-5′-triphosphate, ATP^{2-} as their n-tetrabutylammonium salts by the exo-receptors **1**, **2**, and **3** was carried out by adding the $N(n\text{-}Bu)_4^+$ salt into the electrochemical cells containing a CH_2Cl_2 solution of the exo-receptor at a Pt anode and was efficient, the three oxo-anions giving recognition features that are very different from one another and different from previous dendritic metallocenyl exo-receptors [23, 24, 64, 65, 92]. Selective recognition of ATP^{2-} in the presence of HSO_4^- and Cl^- was also shown. With $H_2PO_4^-$, a progressive wave shift is observed upon titration, and the value of the apparent association constant value using this model is $K_{(+)} = 412 \pm 70$ (SI, Fig. 5).

With adenosine-5′-triphosphate, ATP^{2-}, contrary to the case of $[H_2PO_4]$ $[N(n\text{-}Bu)_4]$, the addition of $[ATP]\,[N(n\text{-}Bu)_4]_2$ into the electrochemical cell containing the CH_2Cl_2 solution of the host provokes the appearance of a new CV wave with these three exo-receptors **1**, **2** and **3**, although each dendrimer shows very different features. In addition, the potential shifts are larger than those observed with $[H_2PO_4]\,[N(n\text{-}Bu)_4]$ (Fig. 11), which is a new situation contrasting to that found with all the metallocenyl dendrimers. The titration diagrams are recorded using the decrease of the intensity of the initial wave and increase of the intensity of the new wave for both **2** and **3**. They show equivalent points for about 0.5 equiv. [ATP] [N(n-Bu)$_4$]$_2$ per Fe$_4$-cluster branch due to the double negative charges, which means that each of the two negative phosphate mono-anion units of ATP^{2-} interacts with a Fe$_4$-cluster branch (Fig. 11). Other stoichiometries have been reported in such titrations [15–18].

The addition of [HSO$_4$][N(n-Bu)$_4$] to the dendrimer **2** provokes the shift of the initial wave that reaches 110 mV at saturation, which leads to an apparent association constant $K_{(+)} = (55 \pm 5)\ 10^3$ L Mol^{-1}. The titration diagram shows an equivalence point at 0.75 equiv. [HSO$_4$] [N(n-Bu)$_4$] per Fe$_4$-cluster branch, although saturation is obtained at 1 equiv. [HSO$_4$] [N(n-Bu)$_4$] per Fe$_4$-cluster branch. In this case the interaction is of the weak type, loose and not selective.

The addition of equimolar amounts of [ATP] [N(n-Bu)$_4$]$_2$, [HSO$_4$] [N(n-Bu)$_4$] and Cl [N(n-Bu)$_4$] to the electrochemical cell containing the dendrimer **2** leads to a shift of the initial wave by 0.1 V. The equivalence point is again reached at 0.5 equiv. [ATP] [N(n-Bu)$_4$]$_2$ which is the signature of the selective recognition of [ATP] [N(n-Bu)$_4$]$_2$, although no new CV wave is observed.

Modification of a Pt electrode with dendrimers such as **2** and **3** is possible (although please note that it cannot be carried out cleanly with the mono-cluster thiol derivative [Fe$_4$(μ_3-CO)$_4$Cp$_3$\{η^5-C$_5$H$_4$CONH(CH$_2$)$_{11}$SH\}], the best results being obtained with **3** due to its larger size ($E_{pa} - E_{pc} = 10$ mV), and recognition of ATP also then proceeds with the replacement of the initial wave by the new wave at less positive potential. After disappearance of the initial wave, the final chemically reversible wave has a $E_{pa} - E_{pc}$ value of 150 mV, signifying that the electron transfer at the electrode surface is slow

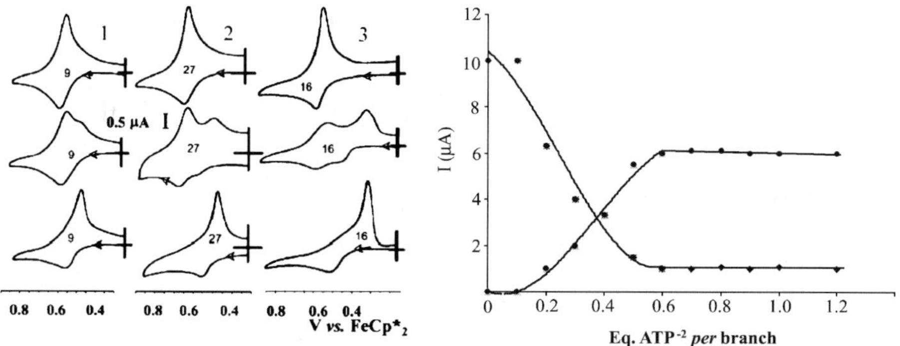

Fig. 11 Comparative titration of ATP^{2-} with **1**, **2** and **3** (6×10^{-5} M) in CH$_2$Cl$_2$: CVs before, during and after addition of [ATP] [N(n-Bu)$_4$]$_2$ (*left*; the number of dendrimer branches are indicated inside the waves). Decrease of the intensity of the initial CV wave (■) and increase of the intensity of the new CV wave (•) *vs.* the number of the equiv. of [(n-Bu$_4$)N]$_2$ [ATP] added per cluster branch of **2** (*right*)

due to structural reorganization of the dendritic host-guest supramolecular assembly that involves, as in solution, formation vs. disruption of large ion pairs in synergy with double hydrogen bonding between the oxo-anion and the amido group [8–14]. The shape of this wave is a fingerprint of the oxo-anion, and [ATP] [N(n-Bu)$_4$]$_2$ can be washed away using CH_2Cl_2 leaving the modified electrode that can be used again.

6
Dendritic Cationic Aminoarene Iron-sandwich Complexes Selectively Recognize Chloride and Bromide Anions

With the dendrimers series shown in Fig. 1, the order of $\Delta E°$ values was: $H_2PO_4^- > HSO_4^- > Cl^- > NO_3^-$. This order can, however, vary greatly from one dendrimer series to the other, and the same can be said of the dendritic effect; some dendrimer families (such as the above one) give a strong dendritic effect for a given anion, and others do not. For instance, an amido-ferrocenyl dendrimer with an octabenzylated durene core and 24 redox termini underwent values of the order of what was found for the dendrimer 9-Fc of Fig. 1, far less than 18-Fc of Fig. 1.

The octasubstituted cored dendrimer 24-FeAr containing 24 cationic aminoarene iron moieties [dendr-(η^6-NHC$_6$H$_4$MeFeCp*)$_{24}$] (Cp* = η^5-C$_5$Me$_5$), i.e. eight tripods, was compared to the analogous mono- and trimetallic amino-arene cationic complexes (resp. 1-FeAr and 3-FeAr, Scheme 9 and Fig. 12) with the same amino-arene iron-sandwich structure for the

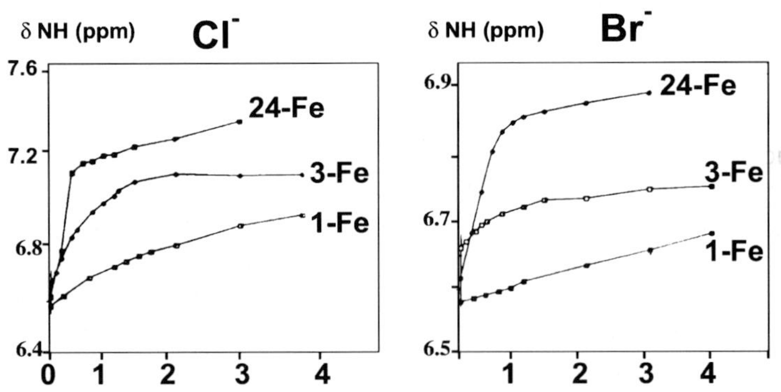

Fig. 12 Variation of δ_{NH} for the exocyclic amine proton of the 24-Fe dendrimer measured by ^{1}H NMR spectroscopy upon addition of n-Bu$_4$NCl or n-Bu$_4$NBr. Comparison with monometallic [FeCp*(η^6-C$_3$H$_5$NHPh)][PF$_6$] (1-Fe) and tripodal PhCH$_2$NHC{(CH$_2$)$_3$O (CH$_2$)$_3$NH-[η^6-PhFeCp*][PF$_6$]}$_3$ (3-Fe) showing the positive dendritic effect

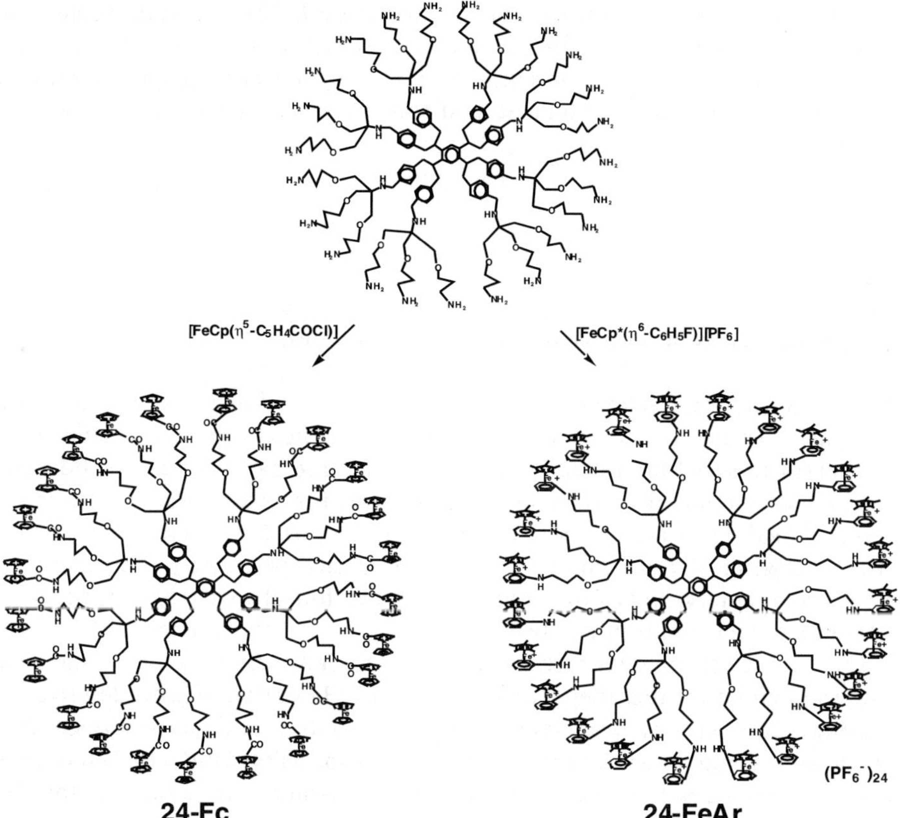

Scheme 9 Syntheses of 24-amidoferrocenyl dendrimer (24-Fc) that recognizes oxo-anions (*left*) and cationic 24-p.aminotolueneFeCp* dendrimer (24-FeAr) that recognizes chloride and bromide (*right*)

recognition of a collection of anions using the NH signal in ^{1}H NMR [102]. The recognition of the oxo-anions that had been successful with the amidoferrocenyl dendrimers was not possible with this system. On the other hand, that of the halides Cl$^-$ and Br$^-$ gave noteworthy results, whereas the 18- and 24-amidoferrocenyl dendrimers hardly gave any significant $\Delta E°$ values for the recognition of these anions. The halides and the secondary amine groups of the dendrimer allow a single hydrogen bonding interaction due to the acidic amino group in contrast to amide groups with oxoanions showing chelating double hydrogen bonding. For bromide, the titration showed an equivalence point also corresponding to a one-to-one interaction, and comparison with the mono- and tripodal compounds containing analogous redox sites indicated that only the dendrimer showed a clear equivalence point. With Cl$^-$, the equivalence point was even sharper and was found for 1/3 of chloride equiv. per branch, i.e. one Cl^- per tripod. This unusual stoichiometry is remarkably specific for Cl^- and shows that only Cl^- fits in the cavity defined by a tripod at the dendrimer periphery [97]. Ferrocenylaminosilane dendrimers reported by Casado et al. also showed recognition properties for the $H_2PO_4^-$ anion [103].

7
Conclusion and Prospects

In this article, we have reviewed anion exo-receptors including metallodendrimers, gold nanoparticles and nanoparticle-cored ferrocenyl dendrimers, most often using cyclovoltammetry that monitors the redox switch. The comparison between mononuclear compounds, tripod derivatives and dendrimers shows that the variation of potential upon titration of anions is larger with dendrimers and increases with the dendrimer generation. This is what was called the positive "dendritic effect" [23, 24] and was accounted for by the narrowing of the channels at the dendrimer periphery, thus allowing a tighter interaction with the anionic guest. This contrasts with the negative dendritic effect found in catalysis whereby the catalytic efficiency decreases upon increasing the dendrimer generation [62, 63]. We have also indicated that the dendritic effect depends on the nature of the anion and dendrimer. Thus amidoferrocenyl dendrimers, gold nanoparticles and gold nanoparticle-cored amidoferrocenyl dendrimers can be excellent sensors with large dendritic effects for oxo-anions, whereas no significant recognition is found for halides whilst the opposite was found with cationic aminoarene-iron-Cp* dendrimers.

Gold nanoparticle-cored ferrocenyl dendrimers are very large (nano-sized) ensembles, containing several hundred equivalent ferrocenyl groups at the dendrimer periphery, that readily adsorb on Pt electrodes, a property that we have used to prepare modified electrodes with these sensors. Selective recognition involving the relatively low-energy hydrogen bonding could be followed by washing of the oxo-anion and re-use of the robust modified electrode several times for further experiments. This contrasts with sensors based on coordination chemistry for which recognition involves the stronger coordination bonds that do not allow re-use of the sensoric units. Finally, the Fe_4-cluster-dendrimers were found to recognize ATP^{2-} better than the model anion $H_2PO_4^-$, apparently due to optimal size matching, which was not the case for the ferrocenyl dendrimers of various types.

Some challenges remain: for instance chiral recognition and simultaneous recognition of both cations and anions [for instance with cobaltocenyl dendrimers [8–14, 104, 105]]; the recognition of biologically important anions in water has still not been successfully achieved despite interesting recent attempts [106, 107]. A higher level of sophistication in the molecular design of the potential receptors and sensors is now clearly required for further progress in these directions. Finally, the advantages of exo-receptors can potentially be applied to sense other substrates such as biologically important amino-acids. Work along these lines is currently underway in our laboratory. Nanoparticles with Fe_4-clusters would allow combination of optimized ATP recognition with the increased adsorption of gold nanoparticles (compared to dendrimers).

Acknowledgements This feature article is dedicated to the unforgettable memory of Professor Olivier Kahn, our distinguished friend and colleague from the University Bordeaux I who passed away in December 1999. The chemistry carried out in our laboratory and reviewed in this article was possible thanks to the ideas and efforts of the students and colleagues cited in the references. Financial support from the Institut Universitaire de France (DA, IUF), the Centre National de la Recherche Scientifique (CNRS), the University Bordeaux I and the Ministère de la Recherche et de la Technologie (M-C D, MRT, thesis grant) is also gratefully acknowledged.

References

1. Czarnik AW (1994) Acc Chem Res 27:302
2. Dürr H, Bouas-Laurent H (ed) (2003) Photochromism. Molecules and Systems, revised ed. Elsevier, Amsterdam
3. Crumbliss AL, Hill HAO, Page D (1988) J Electroanal Chem 206:327
4. Tanigishi I, Miyanoto S, Tomimura S, Hakridge FM (1988) J Electroanal Chem 240:33
5. Heller A (1990) Acc Chem Res 23:228
6. Saiji T, Kinoshita I (1986) J Chem Soc Chem Commun, p 716
7. Edmonds TE (1988) Chemical Sensors, chap. 8. Blackie, Glasgow, p 193
8. Beer PD (1993) Adv Inorg Chem 39:79
9. Beer PD (1996) Chem Commun, p 689
10. Beer PD (1998) Acc Chem Res 31:71
11. Beer PD, Gale PA, Chen Z (1998) Adv Phys Org Chem 31:1
12. Beer PD (2001) Angew Chem Int Ed 40:486
13. Tucker JHR, Collison SR (2002) Chem Soc Rev 31:147
14. Beer PD, Hayes EJ (2003) Coord Chem Rev 240:167
15. Reynes O, Moutet J-C, Pecaut J, Royal G, Saint-Aman E (2000) Chem Eur J 6:2544
16. Reynes O, Royal G, Chainet E, Moutet J-C, Saint-Aman E (2002) Electroanal 15:65
17. Reynes O, Gulon T, Moutet J-C, Royal G, Saint-Aman E (2002) J Organomet Chem 656:116
18. Reynes O, Moutet J-C, Pecaut J, Royal G, Saint-Aman E (2002) New J Chem 26:9
19. Balzani V, Campagna S, Denti G, Juris A, Serroni S, Venturi M (1998) Acc Chem Res 31:26
20. Fillaut J-L, Astruc D (1993) J Chem Soc Chem Commun, p 1320
21. Moulines F, Djakovitch L, Boese R, Gloaguen B, Thiel W, Fillaut J-L, Delville M-H, Astruc D (1993) Angew Chem Int Ed Engl 32:1075
22. Fillaut J-L, Linares L, Astruc D (1994) Angew Chem Int Ed Engl 33:2460
23. Astruc D, Valério C, Fillaut J-L, Ruiz J, Hamon J-R, Varret F (1996) In: Kahn O (ed) Supramolecular Magnetism, NATO ASI Series. Kluwer, Dordrecht, p 107–127

24. Valério C, Fillaut J-L, Ruiz J, Guittard J, Blais J-C, Astruc D (1997) J Am Chem Soc 119:2588
25. Newkome GR, Moorefield CN, Vögtle F (2001) Dendrimers and Dendrons: Concepts, Syntheses, Applications. Wiley, Weinheim
26. Astruc D (ed) (2003) Dendrimers and Nanoscience. CR Chimie 6(8–10), Elsevier, Paris
27. Tomalia DA, Taylor AN, Goddart WA III (1990) Angew Chem Int Ed Engl 29:138
28. Ardoin N, Astruc D (1995) Bull Soc Chem 132:875
29. Moorefield CN, Newkome GN (2003) In: Astruc D (ed) Dendrimers and Nanosciences. CR Chimie, Elsevier, Paris, 6(8–10):715
30. Kaifer AE, Gomez-Kaifer M (1999) Supramolecular Electrochemistry. Wiley, Weinheim, 16:207
31. Kaifer AE, Mendoza S (1996) In: Gokel GW (ed) Comprehensive Supramolecular Chemistry. Pergamon, Oxford, vol 1, 19:701
32. Alonso B, Casado CM, Cuadrado I, Moran M, Kaifer AE (2002) Chem Commun, p 1778
33. Astruc D, Daniel M-C, Ruiz J (2004) Chem Commun 2004:2637
34. Cuadrado I, Morán M, Casado CM, Alonso B, Lobete F, Garcia B, Losada J (1996) Organometallics 15:5278
35. Takada K, Diaz DJ, Abruña H, Cuadrado I, Casado CM, Alonso B, Morán M, Losada J (1997) J Am Chem Soc 119:10763
36. Casado CM, Gonzales B, Cuadrado I, Alonso B, Morán M, Losada J (2000) Angew Chem Int Ed 39:2135
37. Gonzales B, Cuadrado I, Casado CM, Alonso B, Pastor C (2000) Organometallics 19:5518
38. Takada K, Diaz DJ, Abruña H, Cuadrado I, Blanca G, Casado CM, Alonso B, Morán M, Losada J (2001) Chem Eur J 7:1109
39. Casado CM, Cuadrado I, Morán M, Alonso B, Barranco M, Losada J (1999) Appl Organomet Chem 14:245
40. Cuadrado I, Morán M, Casado CM, Alonso B, Losada J (1999) Coord Chem Rev 193–195:395
41. Casado CM, Cuadrado I, Morán M, Alonso B, Garcia B, Gonzales B, Losada J (1999) Coord Chem Rev 185–186:53
42. Alonso B, Alonso E, Astruc D, Blais J-C, Djakovitch L, Fillaut J-L, Nlate S, Moulines F, Rigaut S, Ruiz J, Sartor V, Valério C (2002) In: Newkome GR (ed) Advances in Dendritic Macromolecules. Elsevier, Amsterdam, vol 5:89–127
43. Jutzi P, Batz C, Neumann B, Stammler HG (1996) Angew Chem Int Ed Engl 35:2118
44. Descheneaux R, Serrano E, Levelut A-M (1997) Chem Commun, p 1577
45. Achar S, Immoos CE, Hill MG, Catalano VJ (1997) Inorg Chem 36:2314
46. Shu C-F, Shen H-M (1997) J Mater Chem 7:47
47. Köllner C, Pugin B, Togni A (1998) J Am Chem Soc 120:10274
48. Schneider R, Kollner C, Weber I, Togni A (1999) Chem Commun, p 2415
49. Oosterom EG, van Haaren RJ, Reek JNH, Kamer JPC, van Leewen PWNM (1999) Chem Commun, p 1119
50. Dardel B, Descheneaux R, Even M, Serrano E (1999) Macromolecules 32:5193
51. Mosbach M, Schuhman W (1999) Patent N° DE 19917052, CA 133:293175
52. Ipatschi J, Hosseinzadeh R, Schlaf P (1999) Angew Chem Int Ed Engl 38:1658
53. Turrin C-O, Chiffre J, de Montauzon D, Daran J-C, Caminade A-M, Manoury E, Balavoine G, Majoral J-P (2000) Macromolecules 33:7328
54. Yoon HC, Hong M-Y, Kim H-S (2000) Anal Chem 72:4420

55. Yoon HC, Hong M-Y, Kim H-S (2000) Anal Biochem 282:121
56. Valentini M, Pregosin PS, Ruegger H (2000) Organometallics 19:2551
57. Salmon A, Jutzi P (2001) J Organomet Chem 637–639:595
58. Flanagan JB, Margel S, Bard AJ, Anson FC (1978) J Am Chem Soc 100:4248
59. Bard AJ, Faulkner LR (1980) Electrochemical Methods. Wiley, New York
60. Astruc D (1995) Electron Transfer and Radical Processes in Transition-Metal Chemistry, chap 2. Wiley, New York
61. Miller SR, Gustowski DA, Chen Z-h, Gokel GW, Echegoyen L, Kaifer AE (1988) Anal Chem 60:2021
62. Astruc D, Chardac F (2001) Chem Rev 101:2991
63. Méry D, Astruc D (2006) Coord Chem Rev (in press)
64. Ruiz J, Medel MJ, Daniel M-C, Blais J-C, Astruc D (2003) Chem Commun, p 464
65. Daniel M-C, Ruiz J, Blais J-C, Daro N, Astruc D (2003) Chem Eur J 9:4371
66. Zimmerman SC, Zeng F, Reichert DEC, Kolotuchin SV (1996) Science 271:1095
67. Zeng F, Zimmerman SC (1997) Chem Rev 97:1681
68. Bosman AW, Jensen EW, Meijer EW (1999) Chem Rev 99:1665
69. Smith DK, Diederich F (2000) Top Curr Chem 210:183
70. Fréchet JMJ (2002) Proc Nat Acad Sci 99:4782
71. Gibson HW, Yamaguchi NY, Hamilton L, Jones JW (2002) J Am Chem Soc 124:4653
72. Ermer O, Eling A (1994) J Chem Soc Perkin Trans 2:925
73. Melwyn-Hughes EA (1961) Physical Chemistry, 2nd ed. Pergamon, Oxford, p 1060
74. Desiraju GR (1989) Crystal Engineering: The Design of Organic Solids. Elsevier, New York
75. Hanessian S, Simard M, Roelens S (1995) J Am Chem Soc 117:7630
76. Daniel M-C, Ruiz J, Astruc D (2003) J Am Chem Soc 125:150
77. Daniel M-C, Ba F, Ruiz J, Astruc D (2004) Inorg Chem 43:8649
78. Daniel M-C, Astruc D (2004) Chem Rev 104:293
79. Giersig M, Mulvaney P (1993) Langmuir 9:3408
80. Ung T, Liz-Marzan LM, Mulvaney P (2002) Colloids Surf A: Physicochem Eng Asp 202:119
81. Brust M, Walker M, Bethell D, Schiffrin DJ, Whyman RJ (1994) J Chem Soc Chem Commun, p 801
82. Hasan M, Bethell D, Brust M (2003) J Am Chem Soc 125:1132
83. Templeton AC, Wuelfing WP, Murray RW (2000) Acc Chem Res 33:27
84. Shon Y-S, Choo H, Newkome GR (2003) In: Astruc D (ed) Dendrimers and Nanosciences. CR Chimie, Elsevier, Paris, vol 8–10:1009
85. Weber K, Creager SE (1994) Anal Chem 66:3164
86. Weber K, Stockes JJ, Creager SE (1997) J Phys Chem B 101:8286
87. Labande A, Astruc D (2000) Chem Commun, p 1007
88. Labande A, Ruiz J, Astruc D (2002) J Am Chem Soc 124:1782
89. Daniel M-C, Ruiz J, Nlate S, Palumbo J, Blais J-C, Astruc D (2001) Chem Commun, p 2000
90. Wang R, Yang J, Zheng Z, Carducci MD (2001) Angew Chem Int Ed 40:549
91. Kim M-K, Jeon Y-M, Jeon SW, Hong H-J, Park CG, King K (2001) Chem Commun 40:549
92. Daniel M-C, Ruiz J, Nlate S, Blais J-C, Astruc D (2003) J Am Chem Soc 125:2617
93. Abruña HD (1988) In: Stotheim TA (ed) Electroresponsive Molecular and Polymeric Systems. Dekker, New York 1:97
94. Murray RW (1992) In: Murray RW (ed) Molecular Design of Electrode Surfaces. Techniques of Chemistry XII, Wiley, New York, p 1

95. Horikoshi T, Itoh M, Kurihara M, Kubo K, Nishihara H (1999) J Electroanal Chem 473:113
96. Yamada M, Tadera T, Kubo K, Nishihara H (2001) Langmuir 17:2263
97. Nlate S, Ruiz J, Sartor V, Navarro R, Blais J-C, Astruc D (2000) Chem Eur J 6:2544
98. Westmeyer MD, Massa MA, Rauchfuss TB, Wilson SR (1998) J Am Chem Soc 120:114
99. Alonso E, Ruiz J, Astruc D (2001) CR Acad Sci Paris, Sér. IIc, 3:189
100. Ruiz Aranzaes J, Belin C, Astruc D (2006) Angew Chem Int Ed 45:132
101. Ferguson JA, Meyer TJ (1972) J Am Chem Soc 94:3409
102. Valério C, Alonso E, Ruiz J, Blais J-C, Astruc D (1999) Angew Chem Int Ed 38:1747
103. Casado CM, Cuadrado I, Alonso B, Morán M, Losada J (1999) J Electroanal Chem 463:87
104. Alonso E, Valério C, Ruiz J, Astruc D (1997) New J Chem 21:1139
105. Valério C, Ruiz J, Fillaut J-L, Astruc D (1999) CR Acad Sci Paris, Série IIc, p 79
106. Beer PD, Davis J, Drillsma-Millgrom DA, Szemes F (2002) Chem Commun: 1716
107. Reynes O, Bucher C, Moutet J-C, Royal J-C, Saint-Aman G (2004) Chem Commun, p 428

Top Organomet Chem (2006) 20: 149–176
DOI 10.1007/3418_035
© Springer-Verlag Berlin Heidelberg 2006
Published online: 5 July 2006

Hyperbranched Polymers as Platforms for Catalysts

Chakib Hajji · Rainer Haag (✉)

Institut für Chemie und Biochemie, Freie Universität Berlin, Takustr. 3, 14195 Berlin,
Germany
haag@chemie.fu-berlin.de

1
Introduction

Many common polymeric solid-phase supports show some disadvantages for
their application in catalysis, such as heterogeneous reaction conditions and
therefore unadvantageous kinetics, low loading capacities, exacerbated analysis and in some cases problematic mechanical stability. Because of the
heterogeneity of the support, access to the reactive site is often reduced which
results in poor catalysis. Also the structural information regarding this type
of reaction is limited; while the ability of the catalyst to be recycled is easily
assessed, structural analysis of the precatalyst before as well as after recovery (at the end of the reaction) requires specialized techniques for nonsoluble
polymers. In some cases these problems can be overcome by using soluble
polymeric supports, which have been proposed since the early seventies as an
alternative to conventional solid-phase supports [1]. Dendritic polymers, particularly perfect dendrimers and hyperbranched polymers, offer a wide range
of new possibilities (Fig. 1) [2]. The use of hyperbranched polymers instead of
perfect dendrimers is justified by their similar properties, better accessibility
and lower costs, which are all extremely important for large-scale synthesis. In addition, several of these hyperbranched polymers are commercially
available on a kilogram scale [3].

Dendritic polymers feature homogeneous reaction conditions and enable
the application of standard analytical techniques (TLC, IR, NMR, MALDI-
TOF etc.) as well as the orthogonal use of insoluble reagents. One drawback of

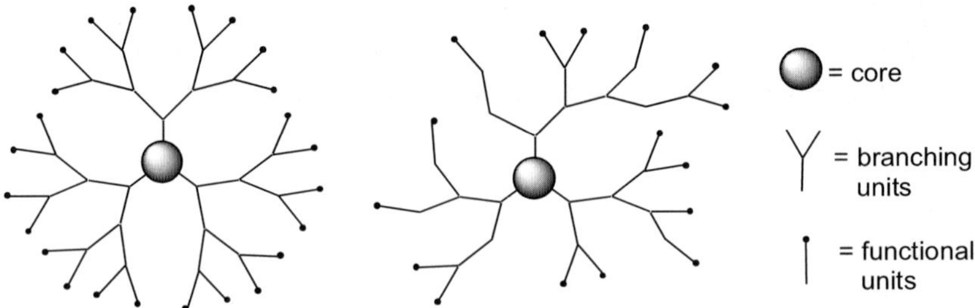

Fig. 1 Schematic comparison of perfect dendrimers and hyperbranched polymers

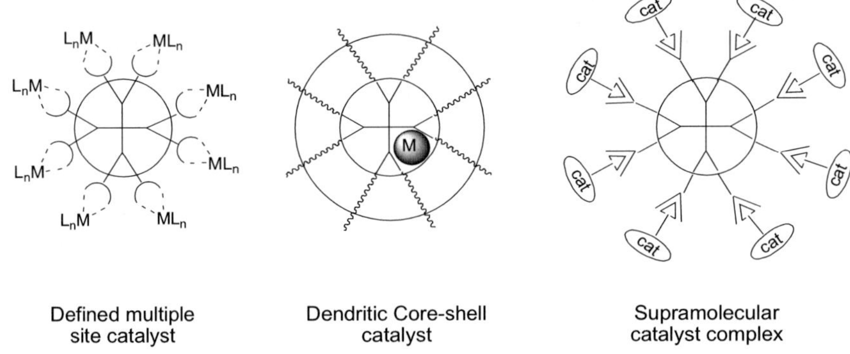

Defined multiple
site catalyst

Dendritic Core-shell
catalyst

Supramolecular
catalyst complex

Fig. 2 Different architectural dendritic catalyst classes discussed in this overview

these soluble supports is the fact that there is no generally applicable separation technique as for solid-phase supports. Depending on the system at hand, several methods of purification are available, such as membrane separation techniques, size exclusion chromatography and precipitation [4].

In this work we present hyperbranched polymers as platforms for catalysts that fall into three major classes, according to their topology and binding mode to the polymeric support (Fig. 2): (i) defined multiple site catalysts; (ii) dendritic core-shell catalysts; (iii) supramolecular catalyst complexes.

2
Defined Multiple Site Catalysts

Dendritic polymers can be covalently functionalized with organometallic complexes to obtain a dendritic catalyst with molecularly defined catalytic sites [5–7]. Moreover, a considerable number of reports on the applicability of functionalized dendrimers in catalysis have led to the idea of a "dendritic effect" on the catalyst activity/selectivity, which can either be positive or negative [8–14]. Since structural perfection may not be a strict prerequisite for well-behaved catalysts, hyperbranched polymers offer a promising alternative for dendrimers in many applications [3, 15–18].

Salazar and coworkers used hyperbranched polyglycerol (PG) amines as ligands in the reaction of oxidative coupling of terminal diacetylenes [19]. For this purpose, they transformed the terminal hydroxyl groups of polyglycerol to the corresponding secondary aliphatic amines (diethylamine or di-n-pentylamine) via tosylates giving amino-terminated hyperbranched molecules $PG-N(Et)_2$ and $PG-N(C_5H_{11})_2$, respectively. These hyperbranched polyamines were then used as ligands in a Cu(I)-catalyzed oxidative coupling reaction of phenylacetylene (Scheme 1).

As a reference, the corresponding monomeric tertiary amines were tested as well as CuCl alone. The results show that the hyperbranched ligands performed better in the oxidative coupling reaction compared to monomeric triethylamine and tripenthylamine (Table 1). This improved performance of

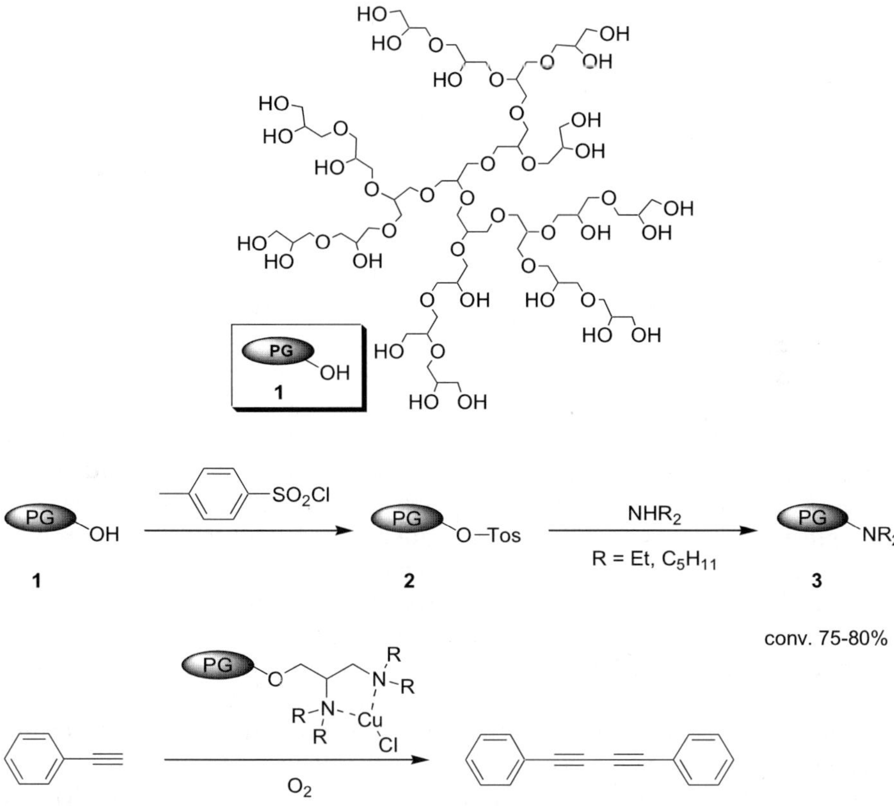

Scheme 1 Structure of hyperbranched polyglycerol and oxidative coupling reaction of phenylacetylene

Table 1 Results of oxidative coupling reaction of phenylacetylene using different amine-CuCl complexes

Amine	Yield (%)
–	< 1
NEt_3	2
$N(n\text{-}C_5H_{11})_3$	< 1
$PG - N(Et)_2$	25
$PG - N(C_5H_{11})_2$	8

the hyperbranched amino ligands can be explained by a better complexation ability and a higher local reagent concentration.

Another application of hyperbranched polymers as supports for catalysts is their use as backbones for the covalent attachment of organometallic fragments. NCN-pincer complexes (NCN-pincer = 2,6-bis[(dimethylamino)-methyl]phenyl anion) are attractive building blocks for catalytic reactions [20, 21]. Covalent introduction of the transition-metal complexes can also be of interest for visualization and imaging of dendritic polymers by transmission electron microscopy (TEM).

For this reason, hyperbranched polyglycerols can be activated by tosylation to introduce platinum NCN-pincer carboxylates by displacement reactions of the tosylate moieties. Stiriba and coworkers reported a synthetic approach using a racemic hyperbranched polyglycerol with the molecular weight M_n = 2000 with trimethylolpropane (TMP) as the initiator. Chiral hyperbranched polyglycerols (–)-PG **1** [M_n = 3000 with bis(2,3-dihydroxypropyl)undecenylamine as initiator] and (+)-PG [M_n = 5500 with trimethylolpropane (TMP) as the initiator] were used as the starting materials. Reaction of tosylated polyglycerols PG – O – Tos **2** with the NCN-pincer platinum carboxylate complex [Pt(NCN – COOK)], obtained by deprotonation of the NCN-pincer platinum compound [Pt(NCN – COOH)], led to NCN-pincer platinum(II)-substituted polyglycerols **4** by nucleophilic displacement of the tosylate groups (Scheme 2). In the substitution reactions, 50% of available tosylate groups was replaced by the organometallic carboxylate in the case of racemic polyglycerol, and only 35% of the available tosylate groups could be replaced in the case of chiral polyglycerols, affording in all cases the modified chiral hyperbranched polyglycerol **4** (Scheme 2). This incomplete substitution was attributed to excessive steric crowding of the relatively bulky pincer system upon higher substitution degrees [22, 23].

The racemic complex **4** containing 50% of NCN-pincer platinum(II) and 50% of tosylate groups has been tested in catalytic double Michael addition of methyl vinyl ketone and ethyl cyanoacetate using (1 mol %) of catalyst.

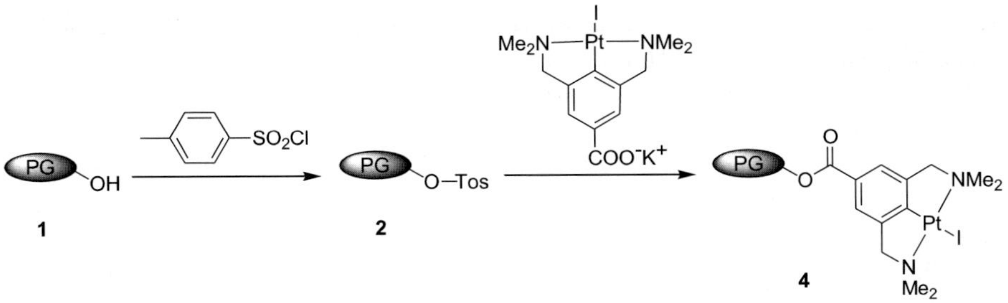

Scheme 2 Synthesis of pertosylated hyperbranched polyglycerol followed by partial substitution of tosyl groups with NCN-pincer platinum(II) carboxylates

Separation of products and catalyst **4** after full conversion was achieved conveniently by dialysis against neat dichloromethane. The catalytic material was recovered in nearly quantitative yields (92%).

To study the effect of the chiral backbone on the behavior of the NCN-pincer complex **4** in catalysis, the Michael addition of methyl vinyl ketone to (R/S)-ethyl α-isocyanopropionate was used as a model reaction (Scheme 3). The results of the experiment show 80% conversion after 24 h which is higher than the 38% conversion obtained in the same Michael addition without catalyst. After full conversion, the product and loaded nanocapsules could be recovered separately in almost quantitative yields (> 96%) by dialysis. Product analysis revealed that no enantiomeric excess was found and racemic mixtures were obtained (ee = 0%) (Scheme 3).

In addition to these noncharged polymer ligands, highly branched polyelectrolytes have also attracted considerable interest [24–29]. This interest stems from the concept of noncovalent binding of catalytically active metal complexes to soluble polymers by electrostatic interactions with the aim of recovering and recycling such catalysts by means of ultrafiltration [4, 30–32]. Cationic hyperbranched polyelectrolytes can be conveniently prepared from polyglycerol in a one-pot reaction with ω-bromoacyl chlorides $Cl(C = O)(CH_2)_n Br$ and 1,2-dimethylimidazole, resulting in a full conversion with respect to esterification and quaternization and affording the final product in 80–90% yield. p-(Diphenylphosphino)benzene sulfonate (TPPMS) was introduced as a counterion into the polyelectrolytes in exchange for the

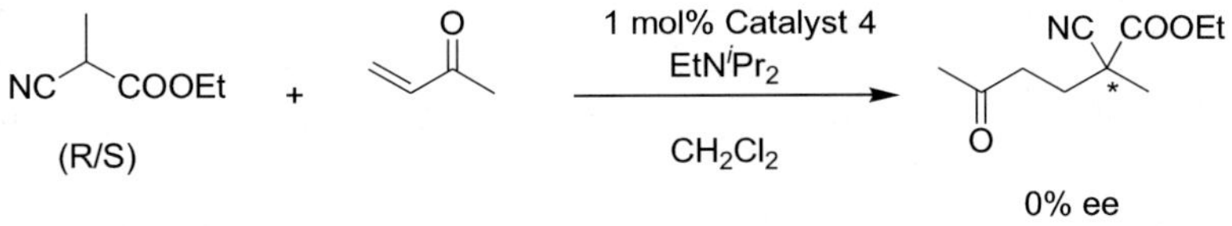

Scheme 3 Catalytic activities of the NCN-pincer complexes in asymmetric Michael addition of methyl vinyl ketone and ethyl α-isocyanopropionate

halides (Scheme 4) [33, 34]. This afforded a polymer with a high number of phosphine functionalities that are able to form catalytically active complexes with transition metals such as rhodium (Fig. 3).

Hydroformylation of 1-hexene was studied at 80 °C under 30 bar of CO/H_2 (1 : 1) with the in-situ generated catalyst **7**. The reaction was monitored via the CO/H_2-uptake which was measured with mass flow meters. The polymer-bound catalysts possessed moderate activities in methanol (200–400 TO/h), and the activity was approx. 4-times higher when non polymer-bound TPPMS was used as the ligand. The activity of the catalyst **7** decreased in recycling experiments (entries 2a–2b in Table 2), which can presumably be attributed to a partial oxidation of the phosphine ligands. Moreover, the activity of the complex was not significantly affected by the change of P : Rh

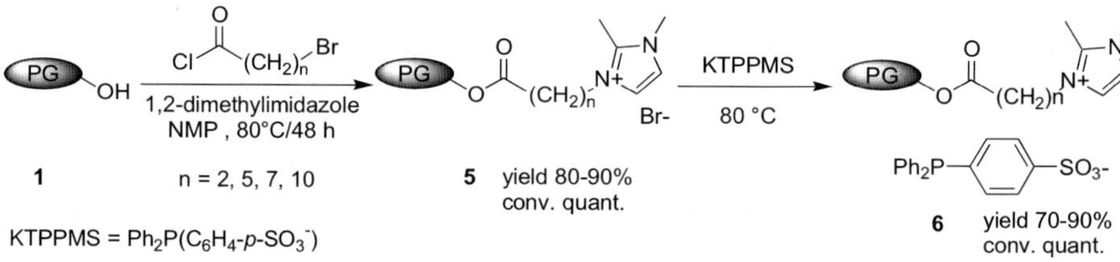

Scheme 4 Synthesis of the hyperbranched imidazolium polycations **6**

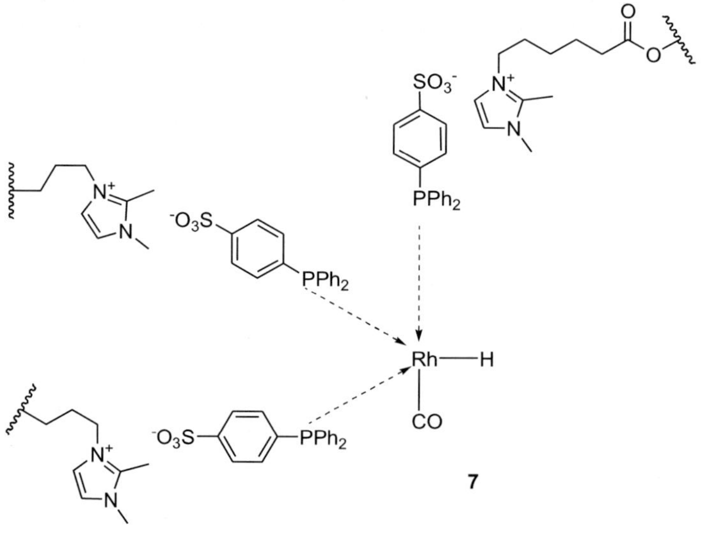

Fig. 3 Schematic representation of the catalyst obtained from the reaction of **6** and $Rh(CO)_2(acac)$ in methanol at elevated temperature under CO/H_2 (1 : 1) pressure (e.g., 80 °C/30 bar) [35]

Table 2 Hydroformylation of 1-hexene with polymer-bound Rh phosphine catalyst 7[a]

Entry	n[b]	Rh (μmol)	P:Rh	1-hexene (mmol)	conv (%)[c]	TO/h	l/b[d]
1		58	10	80	92	1270	3.2
2a	5	37	10	40	33	180	3.6
2b	5	29	10	40	53	360	2.8
3	2	20	10	80	22	390	2.8
4	5	20	10	80	13	260	3.8
5	7	20	10	80	18	320	3.8
6	10	20	10	80	14	275	3.7

[a] Reaction conditions: MeOH, 80 °C; 30 bar CO/H_2 (1 : 1), total reaction volume 100 mL (entry 1 non polymer, for comparison)
[b] Number of carbon atoms of spacer
[c] Conversion to aldehydes determined by GC after 2 h reaction time
[d] linear/branched selectivities

ratio, and the change of flexibility and ion density in the polymer upon variation of the spacer length had no significant influence on the selectivity or activity of the complex (Table 2).

The phenoxazine-based ligand Nixantphos **8**, which was derived from van Leeuwen's initial studies of xanthene-based diphosphine ligands [36], was used as the catalyst in a hydroformylation reaction as well. Nixantphos **8** proved to be a superior ligand with regard to its selectivity towards the linear aldehyde [37]. Moreover, Nixantphos **8** has recently been successfully immobilized on a silica **9** [38–40] and polystyrene **10** [41] matrix. After metal complexation, a catalyst can be obtained that is suitable for recycling by simple filtration (Fig. 4).

Inspired by these developments, Osinski and coworkers employed polyglycerol **1** as the soluble polymer instead of a silica or polystyrene support in order to obtain similarly recyclable, but homogeneous, hydroformylation catalysts [42]. In this context the hydroxyl groups in polyglycerol have been

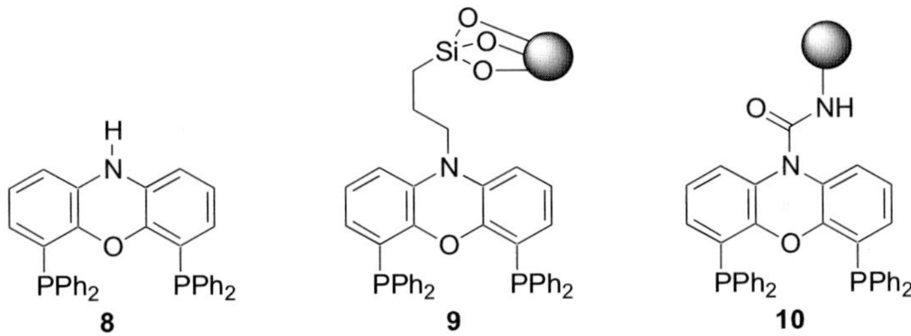

Fig. 4 Solid-phase supported Nixantphos structures

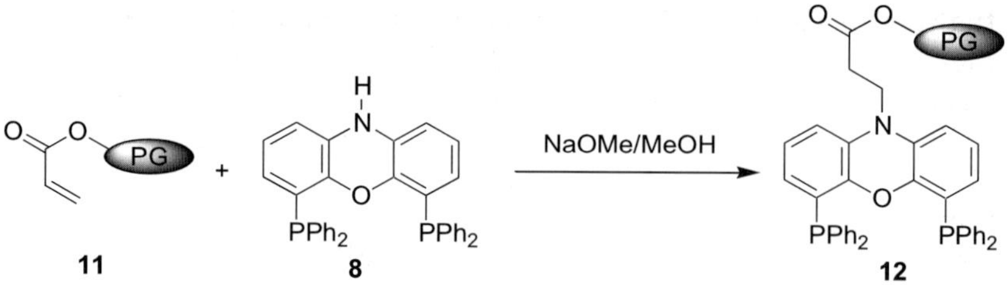

Scheme 5 Attachment of Nixantphos **8** to modified polyglycerol **11**

converted to acryloyl functionality by the reaction of polyglycerol with acryloyl chloride. Michael addition of Nixantphos **8** to this polymer provided
the polyglycerol-supported ligand **12** (Scheme 5). The polymer-supported lig-
and **12** was mixed with rhodium precursors and used as the catalyst in the hydroformylation of *N*-allyl-phthalimide under the same conditions described
in Table 2. In this case, 30–40% conversion and a l:b ratio of 38 : 62 were
observed. The same reaction performed in toluene gave a better conversion
(64%), but exactly the same l:b ratio.

Tetradentate Schiff bases, such as **13** (Fig. 5), also known as salen (*N*,*N'*-
bis-salicylidene-ethylenediamine) are one of the oldest classes of ligands in
coordination chemistry. These systems, e.g., Jacobsen's ligand **14**, are applied as efficient ligands with regard to yield and ee in asymmetric catalysis.
The resulting salen metal complexes and their immobilized analogues have
been used for a wide range of catalytic asymmetric reactions (e.g., asymmetric ring opening of epoxides [43–45], kinetic resolutions of terminal
epoxides [46, 47], and Diels–Alder reactions [48–51]).

Several approaches to immobilize salen systems using polymeric supports
have been reported [52–62]. The rather low conversion could be increased
by improving the loading of the ligand and subsequently the amount of
catalyst per gram of polymer. Among them a highly stereoselective PAMAM-
dendrimer-based salen ligand by Breinbauer et al. was used for hydrolytic
kinetic resolution of epoxides [63].

In principle there are two positions in the ligand structure where appropriate linker units can be attached, either on one of the arene units or

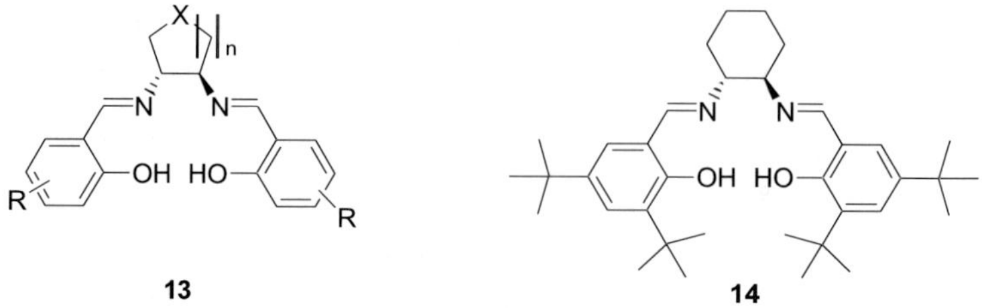

Fig. 5 General structure of salen ligand **13** and Jacobsen's ligand **14**

on the central aliphatic diaminocycle. In this context, Hajji and coworkers synthesized polyglycerol-supported salen analogues and investigated their use in asymmetric synthesis [64]. The polyglycerol-supported unsymmetrical salen ligand **17**, was obtained by reaction of polyglycerol-supported *tert*-butyl salicylaldehyde **15** and mono imino ammonium salt **16** in 65% yield (Scheme 6) [65].

For the application of the polyglycerol salen ligand in catalytic reactions, the introduction of the metal in the corresponding salen is necessary. In the case of the polyglycerol-supported salen ligand **17**, both chromium [66, 67] and cobalt [50] were inserted, and the corresponding metal complexes were characterized by IR and inductive coupled plasma/atomic emission spectrometry (ICP-AES). The chromium insertion takes place employing chromium (II) chloride, and subsequent air oxidation delivered the respective chromium (III) salen complex **18** in 51–73% yield, in which 10–100% of the ligand coordinates a chromium according to ICP-AES. The anion exchange in the complex **18** by reaction with AgBF$_4$ gave the complex **19** in 68% yield, the metal loading being the same as in the precursor. A cobalt (III) salen complex with hexafluoroantimonate counteranions **20** was also synthesized in two steps by means of cobalt insertion, by reaction with Co(OAc)$_2$, and a subsequent anion exchange using AgSbF$_6$, which resulted in the Co(III) complex **20** with 38% yield and metal loading at 28% (Scheme 7).

Transition-metal complexes with salen ligands can serve as catalysts for a multitude of stereogenic reactions including Diels–Alder reactions. We recently studied the application of polyglycerol-supported salen systems **18**, **19** and **20** as chiral catalysts in the hetero Diels–Alder reaction between

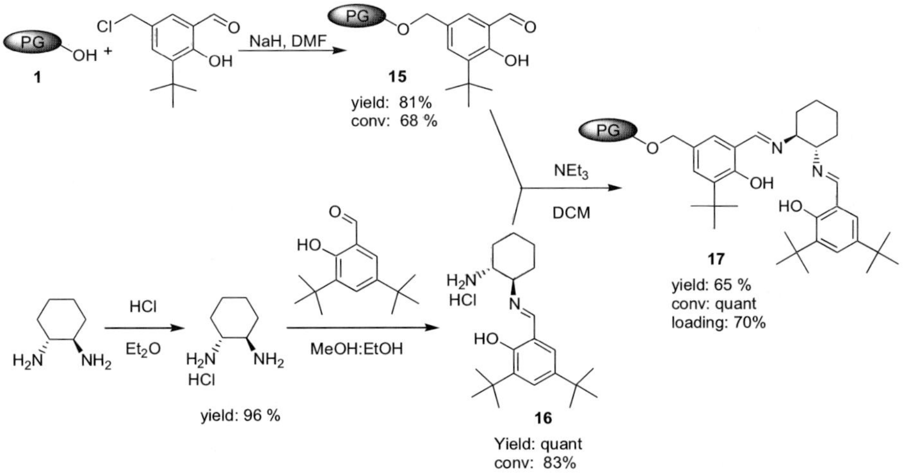

Scheme 6 Synthesis of the polyglycerol-supported unsymmetrical salen ligand **17** from polyglycerol-supported *tert*-butyl salicylaldehyde **15** and monoimine **16**

18: M = Cr, X = Cl   yield 51-73%, conv. 10-100%

19: M = Cr, X = BF$_4$ yield 68%

20: M = Co, X = SbF$_6^-$ yield 38%, conv. 28%

Scheme 7 Introduction of metals (chromium and cobalt) to a polyglycerol-supported salen ligand **17**

trans-1-methoxy-3-trimethylsiloxy-1,3-butadiene (Danishefsky's diene) and benzaldehyde [48, 67, 68]. The results are summarized in Table 3. The activity of the PG-supported catalyst **18** seems to be slightly higher than that of the free Jacobsen's catalyst (entry 3, Table 3).

One of the best-studied examples of recoverable metallodendritic catalyst was reported in 1997 by Reetz for the Heck reaction using Pd dendrimers that were derived from the dendritic phosphines DAB-dendr-[N(CH$_2$PPh$_2$)$_2$]$_x$ ob-

Table 3 Results of Diels–Alder reaction between Danishefsky's diene and benzaldehyde catalyzed by **18**, **19**, and **20** and their comparison with Jacobsen's catalyst **14**-Cr/Cl

Danishefsky's diene

21
(*R*)-2-phenyl-2,3-dihydro-4*H*-pyran-4-one

Entry	Catalyst	Cat. Amount [mol-%]	Temp (°C)[a]	Conv.(%)[b]	ee (%)
1	PG – Cr/Cl **18**	0.5	1[a]	55	64
2	PG – Cr/Cl **18**	1.9	4	64	64
3	Jacobsen's Cr/Cl	1.9	4	23	86
4	–	–	10[a]	< 5	n.d.
5	PG – Cr/BF$_4$ **19**	2.5	–10[a]	13	78
6	PG – Co/SbF$_6$ **20**	1.4	–21[a]	27	63

[a] Temperature at the start of activity.
[b] According to the ^{1}H NMR of the crude product after catalyst removal.

tained by double phosphinomethylation of the commercial DSM polyamino dendrimers DAB-dendr-$(NH_2)_x$ ($x = 4$, 8 or 16 for generations 1, 2 or 3, respectively) [13, 69]. The significant advantages of the soluble dendritic polymeric support were an increased activity, a higher stability and good recyclability compared to the soluble PG-analogue. A few years later, Heuzé reported three generations of new palladodendritic complexes **23a–f** based on dendritic phosphine ligands **22a–f** [14] closely related to Reetz's dendrimers that serve as copper-free recoverable catalysts for the Sonogashira coupling between aryl halides and alkynes. These two families of palladodendrimers (Scheme 8, R = Cy vs. *t*-Bu) show impressively distinct reactivities and recoverabilities, but a rather similar dendritic effect in the Sonogashira reaction.

Generally, the Sonogashira coupling reaction is achieved by a palladium-copper catalyzed reaction of aryl or vinyl halide and terminal alkyne [70–72]. The presence of the copper co-catalyst is an obstacle, however, towards the metallodendritic approach of the system. In this context, only a few examples of copper-free procedures have been reported [73–77], involving for instance, in situ Pd(0) complex formation with bulky phosphines [78].

The dendrimers **23a–c** were used in a copper-free Sonogashira-type coupling reaction between phenylacetylene and iodobenzene or bromobenzene. The catalyst amount was 1 mol % per catalytic group (i.e., 1/4, 1/8 and 1/16 mol % depending on the dendrimer generation; generations 1, 2 and 3), and the temperature range was 25–120 °C.

The results depicted in Table 4 demonstrate the dendritic effect and strong influence of the nature of the diphosphine substituents on the reactivity and recoverability of the catalysts.

Moreover, the metallodendritic catalysts can be recovered after the reaction, which was indeed achieved by precipitation of **23b** and **23c** using pentane. However, the catalysts **23d–f** are too soluble in pentane and other common solvents for recovery because of the presence of the *t*-Bu substituents.

Recently, we used hyperbranched polyglycerol **1** as a high-loading support for boronic acids with application in the palladium-catalyzed Suzuki cross-coupling reaction [79]. Moreover, hyperbranched polyglycerol (**1**) can

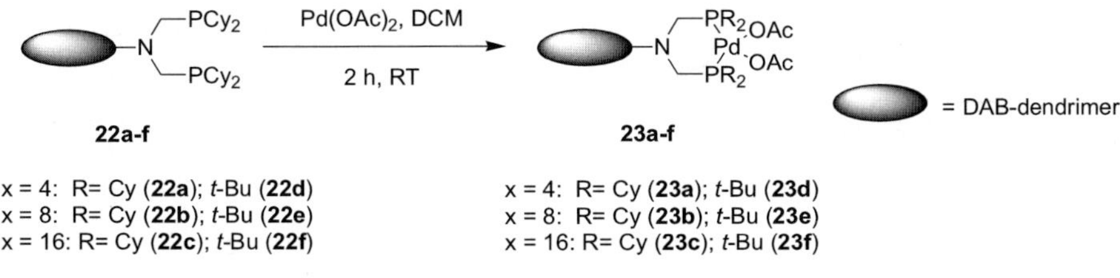

x = 4: R= Cy (**22a**); *t*-Bu (**22d**)   x = 4: R= Cy (**23a**); *t*-Bu (**23d**)
x = 8: R= Cy (**22b**); *t*-Bu (**22e**)   x = 8: R= Cy (**23b**); *t*-Bu (**23e**)
x = 16: R= Cy (**22c**); *t*-Bu (**22f**)   x = 16: R= Cy (**23c**); *t*-Bu (**23f**)

Scheme 8

Table 4 Sonogashira coupling of aryl halide substrates with phenyl acetylene[a]

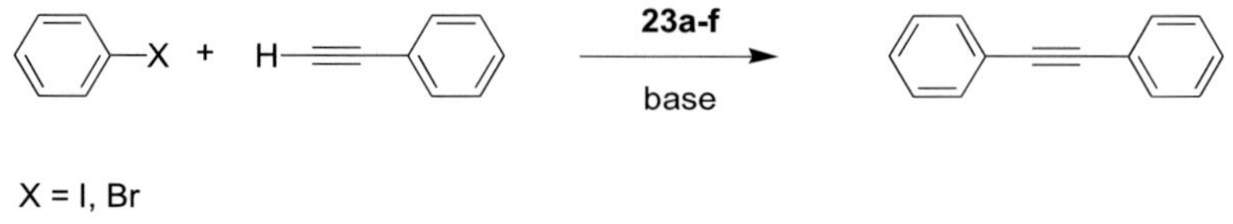

X = I, Br

Catalyst	Aryl halide	Solvent	Temp. (°C)	React. Time (h)	Conv. (%) [b]
23a	Iodobenzene	Et_3N	80	24	79
23b	Iodobenzene	Et_3N	80	24	72
23c	Iodobenzene	Et_3N	80	24	46
23d	Iodobenzene	Et_3N	25	15	97
23e	Iodobenzene	Et_3N	25	40	100
23f	Iodobenzene	Et_3N	25	48	100
23a	Bromobenzene	Et_3N	80	48	17
23b	Bromobenzene	Et_3N	80	48	15
23b	Bromobenzene	Bu_2NH	120	20	20
23c	Bromobenzene	Et_3N	80	48	6
23d	Bromobenzene	Et_3N	25	17	100
23e	Bromobenzene	Et_3N	25	48	93
23f	Bromobenzene	Et_3N	25	48	96

[a] Reaction conditions: aryl halide (2 mmol), phenylacetylene (3 mmol), solvent (6 mL), catalyst (1 mol %), N_2.
[b] The product was isolated by column chromatography.

also be used as the catalyst in the palladium-catalyzed Suzuki coupling reaction between boronic acids and aryl bromides or chlorides. For this purpose, we reported the synthesis of the palladohyperbranched complex **25** based on triphenylphosphino polyglycerol **24** (Scheme 9) [80]. This complex was used as the catalyst in the Suzuki coupling reaction between phenylboronic acid and 4-(bromophenyl)ethanone (Scheme 9). The results of the coupling reaction using palladohyperbranched complex **25** are compared to $Pd(dba)_2$ and summarized in Fig. 6. The catalytic activity and the conver-

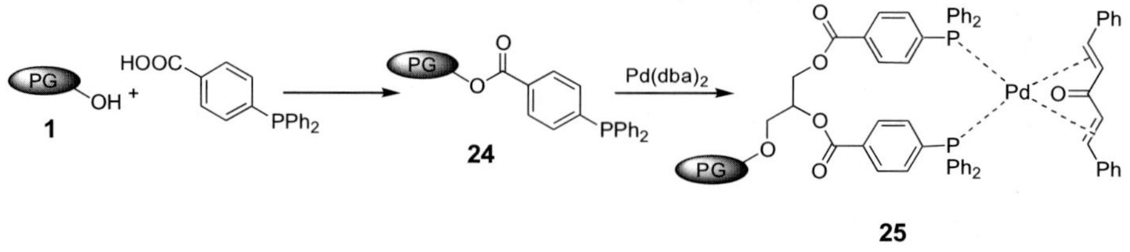

Scheme 9

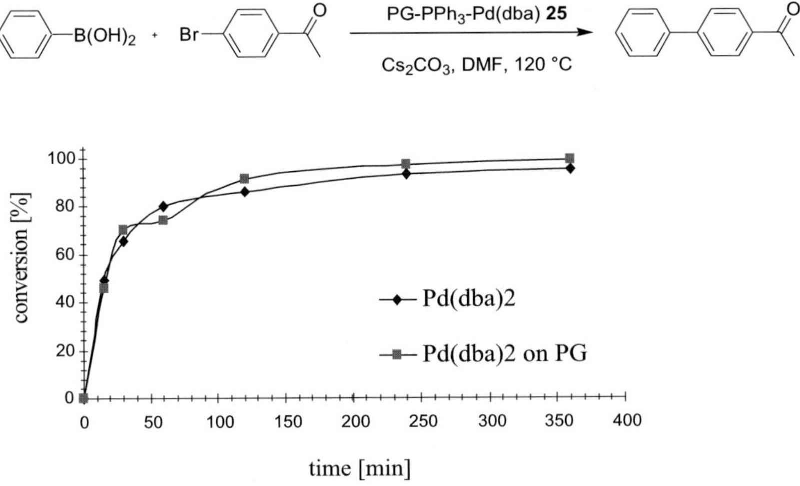

Fig. 6 Comparison of Pd(dba)$_2$ and Pd(dba)$_2$ on PG **25** in Suzuki couplings

sion of the polyglycerol-based Pd-catalyst **25** is comparable to that of the free Pd(dba)$_2$. This again demonstrates that hyperbranched polymers are equally well suited as perfect dendrimers for the covalent attachment of defined multisite catalysts.

3
Dendritic Core-shell Catalysts

Krämer and coworkers recently reported on water-soluble dendritic core-shell architectures and studied the influence of the attached carbohydrate shell on the formation and stabilization of metal nanoparticles in water. For this purpose, they used hyperbranched poly(ethylenimine) (PEI) as core molecules and covalently attached different carbohydrates as shells, i.e., glycidol, gluconolactone and lactobionic acid, to obtain the corresponding PEI-glycol, PEI-gluconamide and PEI-lactobionamide. Different molecular weights of PEI$_x$ ($x = 0.8$, 5, 21 or 25 with different $M_w = \times 10^3$) were employed [81].

Functionalization with glycidol leading to PEI$_{25}$GLY (PEI-glycol, Scheme 10a) could be achieved via addition of DL-glycidol to a solution of PEI$_{25}$, which resulted in a complete conversion of the reactive epoxide group. The attachment of D-glucono-1,5-lactone to PEI$_x$ ($x = 0.8$, 5, 21 or 25), led to PEI$_x$GLU (PEI-gluconamide, Scheme 10b) [82, 83]. The functionalization of PEI$_{25}$ with lactobionic acid gave PEI$_{25}$LAC (PEI-Lactobionamide, Scheme 10c) with 1-ethyl-3-[3-(dimethylamino)propyl]carbodiimide hydrochloride (EDC) used as the coupling reagent in water. Functionalization of the

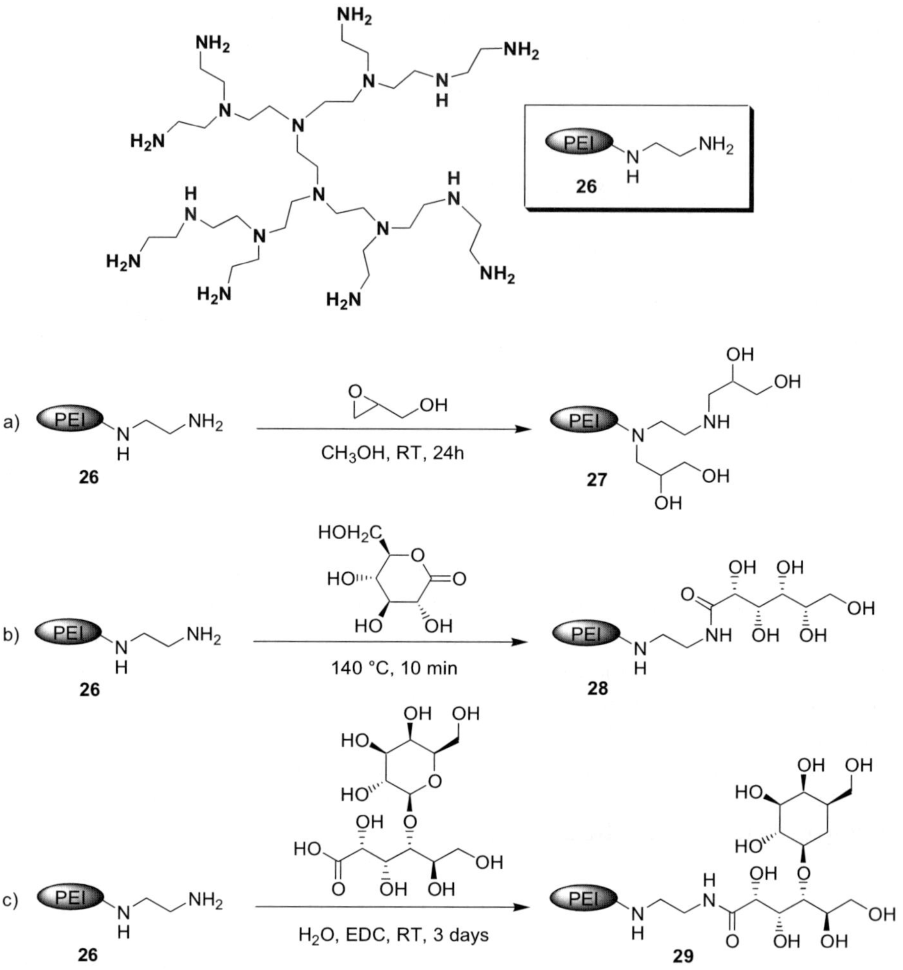

Scheme 10 Structure of hyperbranched poly(ethylene imine) **26** and functionalization of PEI with **a** glycidol (PEI$_x$GLY), **b** gluconic acid (PEI$_x$GLU) and **c** lactobionic acid (PEI$_x$LAC)

linear and terminal monomer units of PEI with a carbohydrate shell yielded a 53% degree of functionalization (DF) in the reaction with glycidol, 33% with gluconlactone and 27% with lactobionic acid.

Hyperbranched PEI or functionalized PEIs with glycidol (PEI-GLY), gluconolactone (PEI-GLU) or lactobionic acid (PEI-LAC) have been used as support materials for metal nanoparticles in water. Polymer-stabilized metal nanoparticles were prepared in a two-step process. After complexation of the metal ions with the respective polymer in a first step, a chemical reduction with sodium borohydride was performed in a second step to obtain the metal nanoparticles. Various metal precursors ($HAuCl_4$, $CuSO_4$ and H_2PtCl_6) were used for the loading of PEIs, and the maximum number of loaded metal ions was found to be around 40 for PEI_5 and PEI_5GLU and is close to the number evaluated for a fourth-generation PAMAM dendrimer (i.e., 50), whereas the molecular weight of the hyperbranched polymer is only half of this.

Various parameters such as pH [84, 85], concentration [86–89], [metal ions]/[polymer] ratio [84, 90] have been optimized to obtain stable nanoparticle systems [81]. The studies show that the dendritic architectures can be used as templates to obtain nanoparticles of defined size. In view of the observed cluster sizes and roughly estimating that the PEI_xGLU or PEI_x gyration radius is around 4.5 nm, a diameter of about 13 nm can only be explained if more than one macromolecule stabilizes a nanoparticle as suggested in the literature in the case of gold nanoparticles stabilized by low-generation PAMAM dendrimers [88]. High molecular weight polymers at low [metal ions]/[polymer] ratios lead, however, to the smallest nanoparticles and might be able to stabilize nanoparticles in the interior of the polymer. The studies demonstrate a clear template effect, i.e., the possibility to control the size of nanoparticles just by changing specific parameters: (i) molecular weight and functionalization and/or (ii) the [metal ions]/[polymer] ratio.

The results obtained from the analysis of TEM measurements show that the functionalization of the amino groups with carbohydrate chains leads to nanoparticles with narrower size distribution and smaller mean particle diameter than for PEI (Figs. 7 and 8). In this case, the resulting particle sizes are comparable to the ones obtained with PAMAM dendrimers [89, 91, 92].

For the applicability of these polymer-stabilized water-soluble nanoparticles in catalysis, the hydrogenation of olefins (i.e., isophorone) with the hyperbranched PEI-GLU-encapsulated platinum nanoparticles was investi-

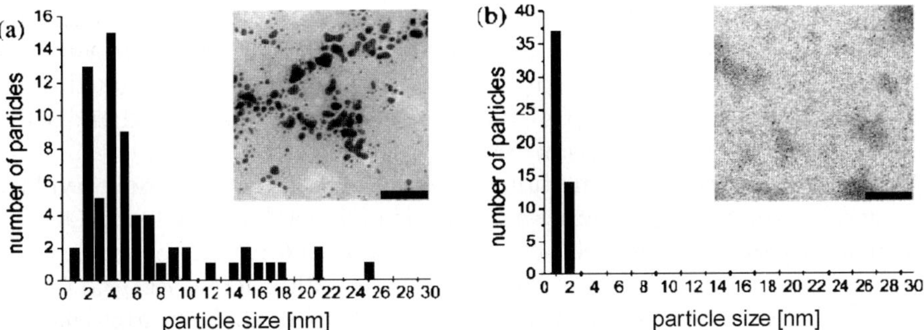

Fig. 7 TEM histograms of Au particles of (**a**) PEI_{21} and (**b**) $PEI_{21}GLU$ (without filtration, partial precipitation occurred after 1 week after reduction); mean diameter of PEI_{21} and $PEI_{21}GLU$ is 6.3 and 1.4 nm, respectively, at a ratio of $[Au^{III}]$/[polymer] of 5. The *black bar* in the TEM pictures corresponds to 50 nm

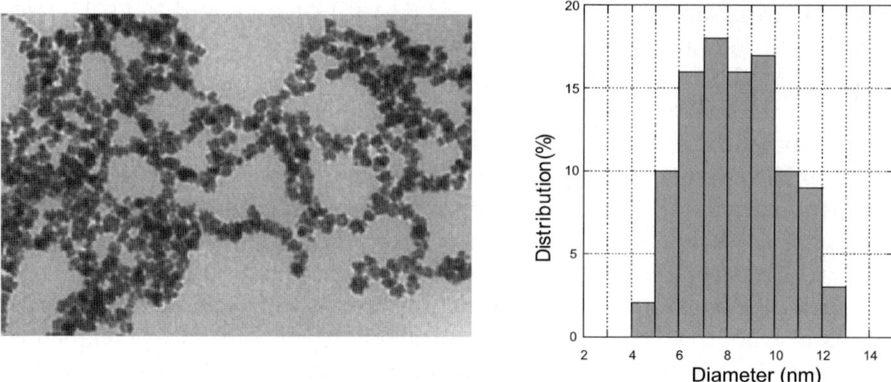

Fig. 8 TEM micrograph and histogram (obtained from about 300 nanoparticles) illustrating the particle size distribution of PEIG-stabilized platinum nanoparticles

Scheme 11 Hydrogenation reaction of isophorone catalyzed by stabilized platinum PEI-GLU nanoparticles

gated in water under mild conditions (Scheme 11, low pressure of H_2 = 2 bar, room temperature).

The conversion rate was measured as well as the turnover number (TONs) defined as the moles of olefin consumed by moles of Pt and the turnover frequencies (TOFs) defined as the moles of isophorone consumed by moles of Pt and by hour. The hydrogenation activities of nanoparticles stabilized by PEI_5, PEI_{25}, PEI_5GLU, $PEI_{25}GLU$, $PEI_{25}GLY$ and $PEI_{25}LAC$ are given in Table 4 as well as the corresponding data for PAMAM in the third and fourth generations ([G3]-PAMAM and [G4]-PAMAM, respectively) measured under the same conditions for comparison (Table 5).

No reduction of the ketone function was observed, and the only product resulting from hydrogenation was 3,3,5-trimethylcyclohexan-1-one. As expected, no significant catalytic activity was observed for Pt-nanoparticles that are not stable on the time scale of the experiment (PEI_5, PEI_{25}, $PEI_{25}GLY$); existence of large clusters not dispersed in aqueous solutions induced a decrease of the Pt surface accessible to the substrate and therefore poor catalytic activity. The catalytic activities of PEI_5GLU were comparable to [G3]-PAMAM- and [G4]-PAMAM-encapsulated platinum nanoclusters and demonstrate that hyperbranched polymers may be more active than perfect dendrimers [93].

Table 5 Results concerning hydrogenation of isophorone catalyzed by platinum-stabilized nanoparticles

	$[Pt^{IV}]/[polym]$	Stability	Conv. [%]	TON[a]	TOF[a,b]
[G3]-PAMAM	20	stable	46.0	23.0	2.9
[G4]-PAMAM	20	stable	50.0	25.0	3.1
PEI_5	20	partial precipitation	0.2	0.1	< 0.1
PEI_5GLU	20	stable	64.0	32.0	4.0
PEI_{25}	20	partial precipitation	0.2	0.1	< 0.1
$PEI_{25}GLU$	20	stable	0.6	0.3	< 0.1
$PEI_{25}GLY$	20	partial precipitation	0.2	0.1	< 0.1
$PEI_{25}LAC$	20	stable	17.0	8.5	0.4

[a] Calculated with respect to the entire amount of Pt present in the clusters, i.e., activity of the significant surface atoms will be higher.
[b] Moles of isophorone consumed by moles of Pt and by hour [mol of isophorone (mol of Pt)$^{-1}$ h^{-1}].

Amphiphilic modification of hyperbranched polyglycerol (**1**) or poly-(ethylene imine) by partial esterification or, respectively, amidation with fatty acids results in macromolecules with a nonpolar shell and a polar core. Metal nanoparticles are stabilized efficiently by these materials, because the polar moieties are responsible for adsorption to the particle surface and the nonpolar region also convey solubility in nonpolar organic solvents [17, 94, 95].

In Sect. 2 of this overview, the NCN-pincer platinum(II) complex was covalently attached to hyperbranched polyglycerol by substitution of tosylate groups to obtain catalyst **4**. This NCN-pincer platinum complex may also be noncovalently immobilized on polyglycerols, and the activity/selectivity of these systems in catalytic reactions has been investigated.

In this context, the chiral hyperbranched polyglycerols (–)-PG [$M_n = 3000$, with bis(2,3-dihydroxypropyl)undecenylamine as the initiator] and (+)-PG [$M_n = 5500$, with trimethylolpropane (TMP) as the initiator] were used. Esterification of the hydroxyl groups of these hyperbranched polyglycerols with hydrophobic alkyl chains as palmitoyl chloride, yielded amphiphilic molecular nanocapsules with reverse micelle-type architecture, in which approximately 50% of the hydroxyl groups were functionalized with palmitoyl chains [96–98]. These materials exhibit low polydispersity ($M_w/M_n < 2$), and the amphiphilic molecular nanocapsules are soluble in nonpolar solvents and irreversibly encapsulate various polar, water-soluble dye molecules in their hydrophilic interior by liquid–liquid extraction [96, 98].

The hydrophilic sulfonated NCN-pincer platinum complex [PtCl(NCN – SO_3H)] was encapsulated by liquid–liquid extractions from an aqueous solution (0.5 M NaOH) into a dichloromethane solution of nanocapsules

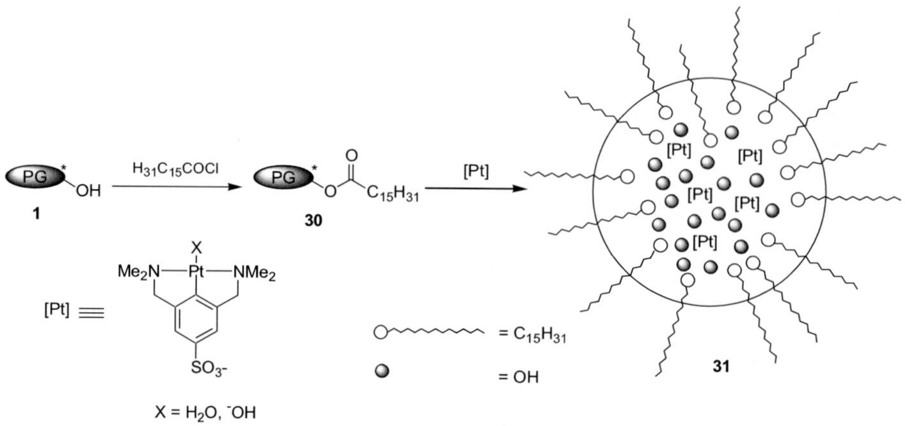

Scheme 12 Preparation of chiral molecular nanocapsules $(-)$-PG-OC(O)$C_{15}H_{31}$ and $(+)$-PG-OC(O)$C_{15}H_{31}$, and a schematic encapsulation of sulfonated platinum pincer complexes [PtCl(NCN$-$SO$_3$H)] within the hydrophilic compartment of the nanocapsules

$(-)$-PG$-$OC(O)$C_{15}H_{31}$ and $(+)$-PG$-$OC(O)$C_{15}H_{31}$, respectively, resulting in the respective chiral nanocapsules loaded with NCN-pincer complex (Scheme 12).

The resulting chiral organometallic materials were used as catalysts in the Michael addition reaction of methyl vinyl ketone to (R/S)-ethyl α-isocyanopropionate, and after 24 h a conversion of 63% was obtained as compared to 80% conversion observed for the covalently immobilized [PtI(NCN–COOK)]. As for the latter, no chirality transfer was observed (see Sect. 2).

In the catalyzed double Michael addition a conversion of 81–95% was observed, as opposed to 40–45% after 40 h in the absence of the pincer complex **31** (Scheme 13) [99].

Moreover, the amphiphilic nonchiral polymer PG$-$OC(O)$C_{15}H_{31}$ was used by Mecking and coworkers in the preparation of nanometer-sized stable palladium colloids using PdCl$_2$ or Pd(OAc)$_2$. The formation of these palladium colloids was visualized by transmission electron microscopy and in the case of the amphiphilic polymer prepared from a PG scaffold of DP$_n$ = 63, a colloid of 5.2 ± 1.8 nm average particle size was obtained. The

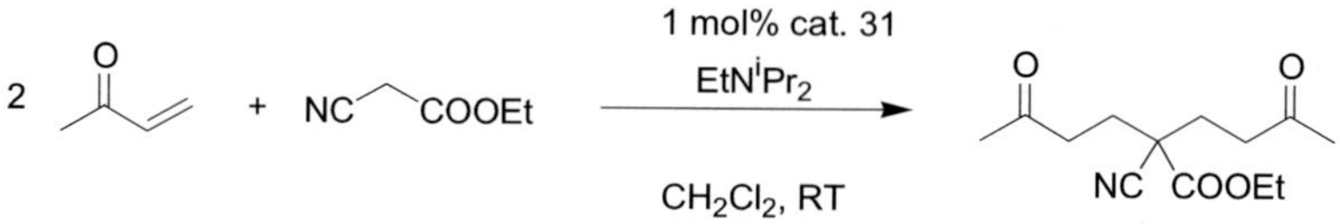

Scheme 13 Application of noncovalent encapsulated platinum pincer complexes in catalytic double Michael addition

metal cluster size decreased to 2.1 ± 0.6 nm for the smaller amphiphilic polyglycerol ($DP_n = 23$) under identical conditions [100]. Given these observed cluster particle sizes, some of the clusters may be stabilized by more than one amphiphilic polymer molecule in both cases. However, upon reduction of the high metal-to-polymer ratios by a factor of ten, much smaller metal particles were obtained, which approached the limit of resolution (ca. 1 nm) of TEM analysis. For the applicability of these polymer-stabilized colloids in catalysis, the hydrogenation of cyclohexene was investigated. Well-behaved kinetics were observed, a high rate of olefin consumption being sustained throughout the course of the reaction, for 75% conversion $\equiv 22.000$ TO (TO = turnovers, i.e., moles of cyclohexane formed per mole of palladium; calculated with respect to the entire amount of Pd present in the nanoparticles, i.e., the activity of the significant surface atoms will be higher) correspond to an overall average rate of 700 TO h^{-1} atm (H$_2$)$^{-1}$ [17]. The catalyst could be recycled and no significant decrease in activity was observed [101–104].

Moreover, this catalytic reaction could be employed in a continuously operated membrane reactor [105, 106]. A stirred membrane reactor module equipped with a solvent-stable Koch MPF-50 membrane [107] was operated at 40 atm. After exchange of a few reactor volumes a steady conversion is achieved, e.g., 30% cyclohexene conversion for the example shown in Fig. 9 [32], corresponding to a catalytic activity of 1200 TO h^{-1}. Over 30 exchanged reactor volumes, corresponding to a time of operation of 30 h, a productivity of a total of 29 000 turnovers was observed.

Comparing colloids with different average particle sizes, an increase in particle size and thus decrease in the number of surface atoms resulted in a decrease in catalytic activity for a given amount of palladium, as expected (Table 6).

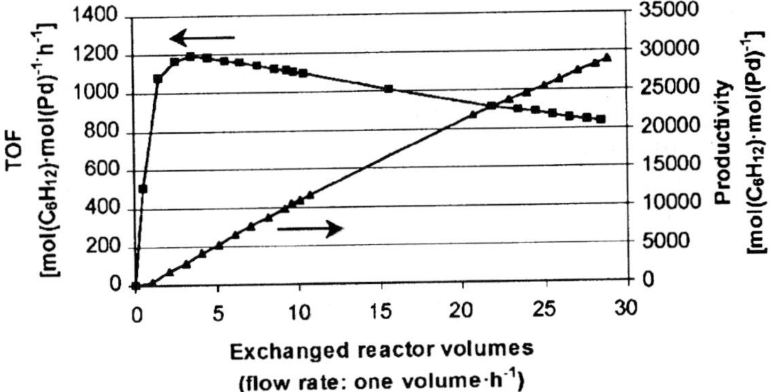

Fig. 9 Hydrogenation of cyclohexene in a continuously operated membrane reactor catalyzed by soluble hybrids of palladium nanoparticles with highly branched amphiphilic polyglycerol. Conditions, see Table 6 (particle size 2.2 ± 0.5 nm)

Table 6 Continuous hydrogenation with colloids of different particle size[a]

Average particle size	Exchanged reactor volumes n (duration of run)	Activity during n^{th} volume [TO h^{-1}][b]
2.2 ± 0.5 nm	4	1190
	10	1100
	25	870
4.9 ± 1.3 nm	4	710
	10	620
	25	350

[a] Temperature: 23 °C, 1.78 µmol Pd in 20 mL reactor volume, flow rate: one reactor volume per hour; feed: 0.33 M in cyclohexene, approx. 0.12 M in H_2.
[b] Activity calculated with respect to the entire amount of palladium present, i.e., activity of the significant surface atoms will be higher

4
Supramolecular Catalyst Complexes

Supramolecular catalyst complexes are yet another class that has been investigated by various groups using different approaches to noncovalent immobilization of the respective ligand on the dendritic polymer. A number of supramolecular complexes have been prepared and tested by Baars and coworkers (Scheme 14) [108, 109]. For application of the resulting homogeneous supramolecular complex in catalytic reactions, different palladium complexes were formed by addition of Pd(COD)MeCl or [(crotyl)PdCl]$_2$ (Scheme 14a). A second approach toward the formation of metal-functionalized systems involved the synthesis of the metal complex of acid **32** prior to the noncovalent anchoring of the complex to the periphery of the dendrimer, leading to the same product (Scheme 14b). Upon the addition of 0.5 equiv of (COD)PdMeCl with respect to the phosphine ligand, *trans* complexes were formed. A third route toward Pd-functionalized dendrimers involved the mixing of all the reactants simultaneously (Scheme 14c). In this way they prepared [{acid(**32**)}Pd(crotyl)Cl]$_{32}$-, and [{acid(**32**)}$_2$Pd(crotyl)Cl]$_{16}$-dendrimer complexes.

A fifth generation poly(propylene imine) dendrimer functionalized with urea adamantyl units at the periphery **33** provided directional binding sites for the strong but reversible binding of 32 guest molecules functionalized with the complementary binding motive. The binding is based on a combination of ionic interactions and the formation of multiple hydrogen bonds. It was previously found that both the urea and the carboxylic acid functionality must be present in the guest molecules to achieve the required strong binding to the dendrimeric host **33** [108, 109].

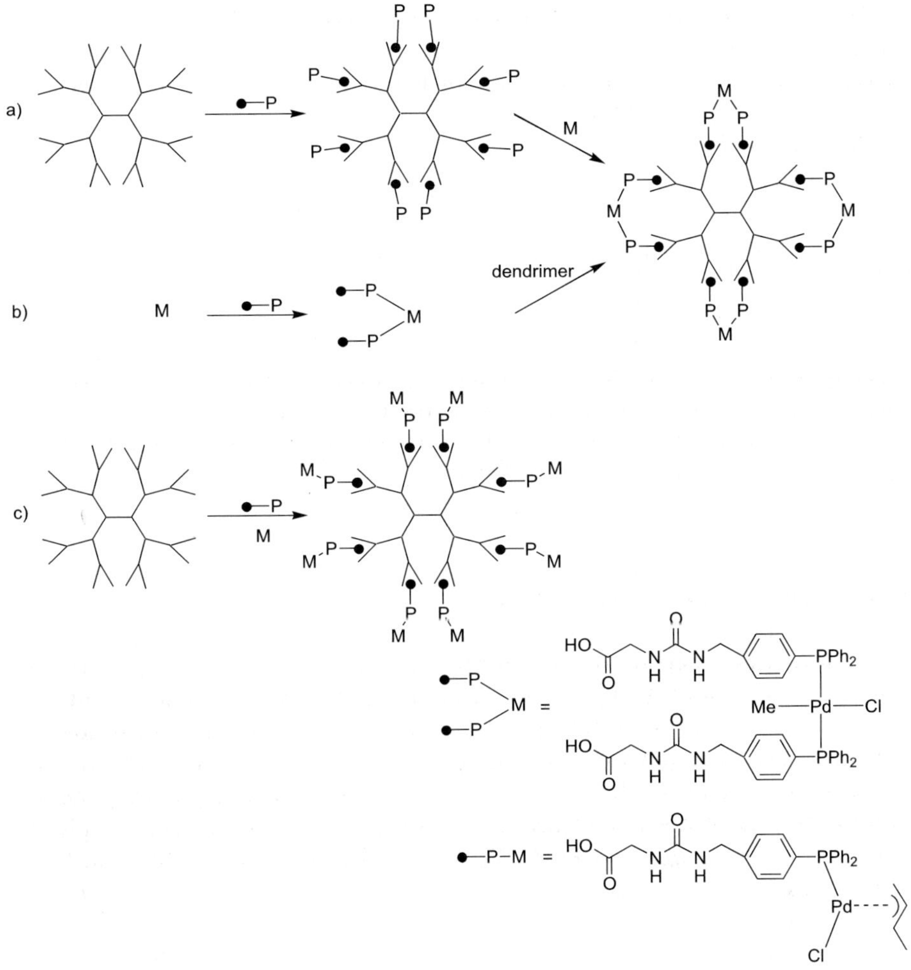

Scheme 14 Schematic representation of the preparation of transition-metal complexes using ligands which are noncovalently anchored to the periphery of a dendrimer

In this case, the phosphine guest **32** was prepared in two simple reaction steps. A reaction of *p*-(diphenylphosphino)benzylamine with commercially available ethyl isocyanatoacetate yielded the ester derivative, followed by hydrolysis of the ester and recrystallization to give a pure **32** [27].

The binding of the guest molecule to the periphery of **33** was studied by NMR spectroscopy, and this shows a shift of the urea protons of the dendrimer from 6.20 and 5.43 ppm to 6.34 and 5.64 ppm, respectively, resulting after the addition of 32 equiv of acid **32** to **33**. This shift is similar to that previously reported for other guest molecules [108]. In a ^{1}H-NOE NMR experiment of the dendrimer with 32 equiv of acid **32** the aromatic protons of the guest were selectively irradiated and a NOE effect was observed for the adamantyl protons. This shows that the diphenylphosphine groups of the guest and the adamantyl end groups of the dendrimer are in close proximity. A 2D-NOESY NMR spectrum also showed the interaction of the aromatic protons of the guest molecule with the protons of the adamantyl groups. These experiments suggest that multiple hydrogen bonds are formed (Fig. 10).

The dendrimers [{acid(32)}$_2$Pd(crotyl)Cl]$_{16}$ (P/Pd = 2) and [{acid-(32)}Pd(crotyl)Cl]$_{32}$ (P/Pd = 1) have been used as catalysts in the allylic amination [110] of crotyl acetate with piperidine (Table 7). These experiments clearly show that the supramolecular anchoring of the catalysts does not decrease the activity, which generally is observed for catalysts immobilized on an insoluble support. The selectivity using the catalyst (P/Pd = 1) is slightly different than obtained with a P/Pd ratio of 2 [111–113]. Moreover, application of the ester complex Pd(crotyl)Cl (M_W = 617.4) in a continuous flow membrane reactor [114–117] led to a retention measured in the presence of the host dendrimer of 97%, which is too low for practical application. In contrast, the supramolecular acid-dendrimer complex [(32)Pd(crotyl)Cl]$_{32}$ had a membrane retention as high as 99.4%. Interestingly, upon using a lower palladium loading (P/Pd = 2) the retention further increased to 99.9%, suggesting that the palladium diphosphine complexes are bound more strongly due to cooperative effects. These results indicate that this type of supramolecular complex will be efficiently separated from smaller compounds such as the reactants and products in a continuous-flow membrane reactor.

Hyperbranched polymers can also be used for supramolecular immobilization (Scheme 15). Yet another approach for the noncovalent immobilization has been presented by Tzschucke and coworkers who used interactions between fluorous phase silica (FPS) and perfluoro-tagged palladium

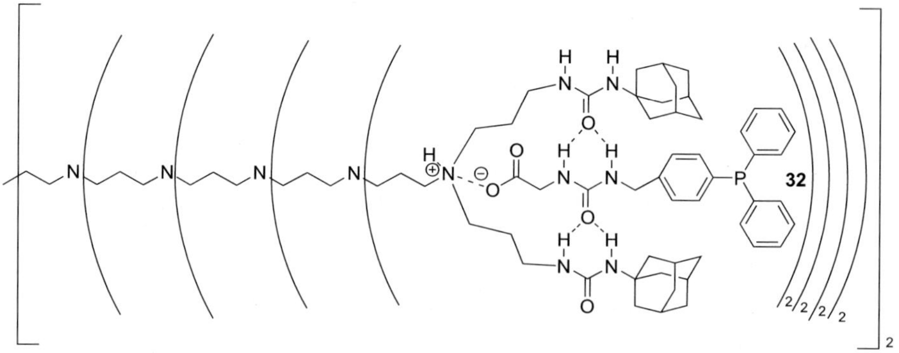

Fig. 10 Phosphine ligands assembled to the periphery of a urea adamantyl-functionalized poly(propylene imine) dendrimer **33**

Table 7 Results of the Pd-catalyzed allylic amination of crotyl acetate and piperidine, comparing the supramolecular dendrimeric catalyst with the corresponding monomer[a]

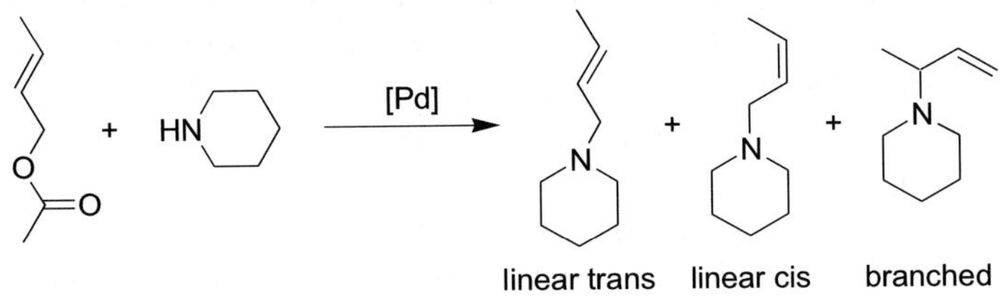

linear trans   linear cis   branched

Catalyst	P/Pd	Conv.[b] (%)	*trans* (%)	*cis* (%)	branched (%)
{ester}$_2$Pd(crotyl)Cl[a]	2	89	38	14	48
{acid(32)}$_2$Pd(crotyl) Cl-dendrimer[a]	2	91	37	15	48
{ester}Pd(crotyl)Cl[c]	1	80	33	6	61
{acid(32)}Pd(crotyl) Cl-dendrimer[c]	1	72	33	6	61

[a] Room temperature; solvent, CH_2Cl_2; volume, 5 mL; [crotyl acetate] = 0.2 M, [piperidine] = 0.4 M, [Pd] = 0.002 M
[b] Conversion after 5 min
[c] Room temperature; solvent, CH_2Cl_2; volume, 5 mL; [crotyl acetate] = 0.12 M, [piperidine] = 0.24 M, [Pd] = 0.004 M

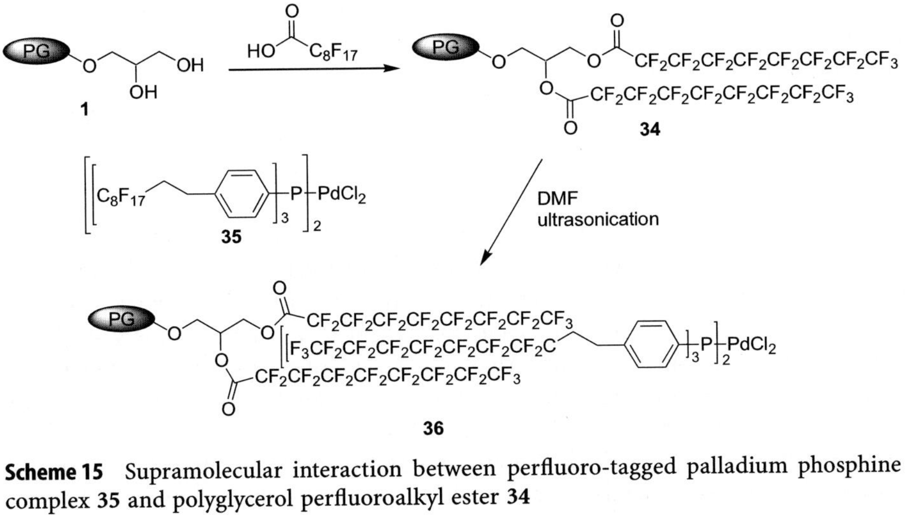

Scheme 15 Supramolecular interaction between perfluoro-tagged palladium phosphine complex **35** and polyglycerol perfluoroalkyl ester **34**

Table 8 Results of consecutive Suzuki couplings with catalyst **36**

Catalyst loading	Yield [%] (1st run)	Yield [%] (2nd run)	Yield [%] (3rd run)	Yield [%] without support **35**
0.1 mol%	> 99	29	3	26
0.5 mol%	> 99	> 99	97	29
1 mol%	> 99	> 99	> 99	55

complexes for the immobilization and recycling of catalysts in organic solvents [111, 118]. Hyperbranched polymers can also be modified covalently with perfluorinated shells, and can encapsulate ions, polar dyes and metal nanoparticles [119]. In this context, Garcia-Bernabé and coworkers immobilized a perfluoro-tagged palladium catalyst on a dendritic polyglycerol ester with a perfluorinated shell and investigated its catalytic activity in Suzuki couplings (Scheme 15) [120].

The immobilized perfluoro-tagged palladium complex **36** is highly active in Suzuki coupling reactions (> 99% yield), and can easily be recycled by precipitation in a mixture of DME/H$_2$O (10% HCl) 2 : 1 and reused for three consecutive Suzuki coupling reactions (Table 8).

This supramolecular dendritic assembly serves as a valuable soluble model for the interaction of perfluoro-tagged catalysts with insoluble supports such as fluorous silica gel and clearly reveals the ligand diffusion from the complex at elevated temperatures. This behavior can also explain the high catalytic activity of the heterogeneous FPS system.

5
Conclusions

The use of hyperbranched polymers instead of perfect dendrimers as platforms for catalysts is justified by their similar homogenous reaction properties, better accessibility and lower costs, which are all extremely important for large-scale applications. Also, several of these hyperbranched polymers are commercially available on a kilogram scale (i.e., polyglycerol **1** and polyethyleneimine **26**). In addition, membrane reactor technology has dramatically improved over the last few years and provides a number of systems with high stability towards organic solvents and reagents. Therefore, hyperbranched polymers provide a great potential for the future as platforms for catalysts.

References

1. Buchmeiser MR (2003) Polymeric Materials in Organic Synthesis and Catalysis. Wiley, New York
2. Haag R (2001) Chem Eur J 7:327
3. Frey H, Haag R (2001) In: Buschow KHJ, Cahn RH, Flemings MC, Ilschner B, Kramer EJ, Majahan S (eds) Encyclopedia of Materials, Science and Technology. Elsevier, Oxford, p 3997
4. Tzschuke CC, Markert C, Bannwarth W, Roller S, Hebel A, Haag R (2002) Angew Chem Int Ed 41:3964
5. Knapen JWJ, van der Made AW, de Wilde JC, van Leeuwuwen PWNM, Wijkens P, Grove DM, van Koten G (1994) Nature 372:659
6. Kleij AW, Kleijn H, Jastrzebski JTBH, Smeets WJJ, Spek AL, van Koten G (1999) Organometallics 18:277
7. Astruc D, Blais J-C, Cloutet E, Djakovitch L, Rigaut S, Ruiz J, Sartor V, Valério C (2000) Top Curr Chem 210:229
8. Brunner H, Fürst J (1994) Tetrahedron 50:4303
9. Brunner H (1995) J Organomet Chem 500:39
10. Bhyrappa P, Young JK, Moore JS, Suslick KS (1996) J Am Chem Soc 118:5708
11. Mack CC, Chow H-F (1997) Macromolecules 30:1228
12. Bhyrappa P, Vaijayanthimala G, Suslick KS (1999) J Am Chem Soc 121:262
13. Gatard S, Nlate S, Cloutet E, Bravic G, Blais J-C, Astruc D (2003) Angew Chem Int Ed 42:452
14. Heuzé K, Méry D, Gauss D, Astruc D (2003) Chem Commun, p 2274
15. Jikei M, Makimoto M (2001) Prog Plym Sci 26:1233
16. Schlenk C, Kleij AW, Frey H, van Koten G (2000) Angew Chem Int Ed 39:3445
17. Mecking S, Thomann R, Frey H, Sunder A (2000) Macromolecules 33:3958
18. Haag R, Roller S (2004) Top Curr Chem 242:1
19. Salazar R, Fomina L, Fomine S (2001) Polymer Bulletin 47:151
20. Albrecht M, van Koten G (2001) Angew Chem Int Ed 40:3750
21. Slagt MQ, Gebbink RJMK, Lutz M, Spek AL, van Koten G (2002) J Chem Soc Dalton Trans, p 2591
22. Stiriba S-E, Slagt MQ, Kautz H, Gebbink RJMK, Thomann R, Frey H, van Koten (2004) Chem Eur J 10:1267
23. Slagt MQ, Stiriba S-E, Kautz H, Gebbink RJMK, Frey H, Van Koten G (2004) Organometallics 23:1525
24. Turner SR, Walter F, Voit BI, Mourey TH (1994) Macromolecules 27:1611
25. Gong A, Liu C, Chen Y, Zhang X, Chen C, Xi F (1999) Macromol Rapid Commun 20:492
26. Van Duijvenbode RC, Rajanayagam A, Koper GJM, Baars MWPL, de Waal BFM, Meijer EW (2000) Macromolecules 33:46
27. De Groot D, de Waal BFM, Reek JNH, Schenning APHJ, Kamer PCJ, Meijer EW, van Leeuwen PWNM (2001) J Am Chem Soc 123:8453
28. Van de Coevering R, Kuil M, Klein Gebbink RJM, van Koten G (2002) Chem Commun, p 1636

29. Mori H, Seng DC, Lechner H, Zhang M, Müller AHE (2002) Macromolecules 35:9270
30. Schwab E, Mecking S (2001) Organometallics 20:5504
31. Mecking S, Thomann R (2000) Adv Mater 12:953
32. Sablong R, Schlotterbeck U, Vogt D, Mecking S (2003) Adv Synth Catal 345:333
33. Schwab E, Mecking S (2005) J Polym Sci Part A Polym Chem 43:4609
34. Schwab E, Mecking S (2005) Organometallics 24:3758
35. Herrmann WA, Kulpe JA, Kellner J, Riepl H, Bahrmann H, Konkol W (1990) Angew Chem Int Ed Engl 29:391
36. van der Veen LA, Boele MDK, Bregman FR, Kamer PCJ, van Leeuwen PWNM, Goubitz K, Fraanje J, Schenk H, Bo C (1998) J Am Chem Soc 120:11616
37. Claver C, van Leeuwen PWNM (eds) (2000) Rhodium Catalyzed Hydroformylation. Kluwer, The Netherlands
38. van Leeuwen PWNM, Sandee AJ, Reek JNH, Kamer PCJ (2002) J Mol Catal A 182–183:107
39. Sandee AJ, Reek JNH, Kamer PCJ, van Leeuwen PWNM (2001) J Am Chem Soc 123:8468
40. Sandee AJ, van der Veen LA, Reek JNH, Kamer PCJ, Lutz M, Spek AL, van Leeuwen PWNM (1999) Angew Chem Int Ed 38:3231
41. Deprèle S, Montchamp J-L (2004) Org Lett 6:3805
42. Ricken S, Osinski PW, Eilbracht P, Haag R (2006) J Mol Cat A (in press)
43. Angelino MD, Laibinis PE (1999) J Polym Sci Part A Polym Chem 37:3888
44. Ready JM, Jacobsen EN (2001) J Am Chem Soc 123:2687
45. Ready JM, Jacobsen EN (2002) Angew Chem Int Ed 41:1374
46. Annis DA, Jacobsen EN (1999) J Am Chem Soc 121:4147
47. Mihara J, Hamada T, Takeda T, Irie R, Katsuki T (1999) Synlett, p 1160
48. Heckel A, Seebach D (2002) Helv Chim Acta 85:913
49. Chapman JJ, Day CS, Welker ME (2001) Eur J Org Chem (12):2273
50. Huang Y, Iwama T, Rawal VH (2002) J Am Chem Soc 124:5950
51. Kwiatkowski P, Asztemborska M, Jurszak J (2004) Tetrahedron: Asymmetry 15:3189
52. Nicolaou KC, Pfefferkorn JA (2001) Biopolymers 60:171
53. Hudson D (1999) J Comb Chem 1:333
54. Barrett AGM, Cramp SM, Roberts RS (1999) Org Lett 1:1083
55. Buchmeiser MR, Wurst K (1999) J Am Chem Soc 121:11101
56. Harned AM, Song He H, Toy PH, Flynn DL, Hanson PR (2005) J Am Chem Soc 127:52
57. Bergbreiter DE (2002) Chem Rev 102:3345
58. Kreiter R, Kleij AW, Gebbink RJMK, van Koten G (2001) In: Vögtle F, Schally CA (eds) Topics in Current Chemistry. Springer, Berlin Heidelberg New York, p 163
59. Eda M, Kurth MJ (2001) Chem Commun, p 723
60. Houlton S (2003) Chemicals Magazine, p 10
61. Rouhi AM (2002) Chem Eng News 80:43
62. Schaus SE, Brandes BD, Larrow JF, Tokunaga M, Hansen KB, Gould AE, Furrow ME, Jacobsen EN (2002) J Am Chem Soc 124:1307
63. Breinbauer R, Jacobsen EN (2000) Angew Chem Int Ed 39:3604
64. Hajji C, Roller S, Beigi M, Liese A, Haag R (2006) Adv Synth Catal (in press)
65. Campbell EJ, Nguyen ST (2001) Tetrahedron Lett 42:1221
66. Martínez LE, Leighton JL, Carsten DH, Jacobsen EN (1995) J Am Chem Soc 117:5897
67. Schaus SE, Branalt J, Jacobsen EN (1998) J Org Chem 63:403
68. Sellner H, Karjalainen JK, Seebach D (2001) Chem Eur J 7:2873
69. Reetz MT, Lohmer G, Schwickardi R (1997) Angew Chem Int Ed Engl 36:1526

70. Sonogashira K, Tohda Y, Hagihara N (1975) Tetrahedron Lett 50:4467
71. Campbell IB (1994) In: Taylor RJK (ed) Organocopper Reagents: A Practical Approach. Oxford University Press, Oxford, UK, p 218
72. Rossi R, Carpita A, Bellina F (1995) Org Prep Proc 27:129
73. Uozumi Y, Kobayashi Y (2003) Heterocycles 59:71
74. Pal M, Parasuraman K, Gupta S, Yeleswarapu KR (2002) Synlett 12:1976
75. Fu X, Zhang S, Yin J, Schumacher DP (2002) Tetrahedron Lett 43:6673
76. Alonso DA, Nájera C, Pacheco MC (2002) Tetrahedron Lett 43:9365
77. Farina V, Kapadia S, Krishnan B, Wang C, Liebeskind SL (1994) J Org Chem 59:5905
78. Böhm VP, Herrmann WA (2000) Eur J Org Chem 22:3679
79. Hebel A, Haag R (2002) J Org Chem 67:9452
80. Roller S, Zhou H, Haag R (2005) Molecular Diversity 9:305
81. Krämer M, Pérignon N, Haag R, Marty J-D, Thomann R, de Viguerie NL, Mingotaud C (2005) Macromolecules 38:8308
82. Schmitzer AR, Perez E, Rico-Lattes I, Lattes A, Rosca S (1999) Langmuir 15:4397
83. Schmitzer AR, Franceschi S, Perez E, Rico-Lattes I, Lattes A, Thion L, Erard M, Vidal C (2001) J Am Chem Soc 123:5956
84. Sidorov SN, Bronstein LM, Valetsky PM, Hartmann J, Cölfen H, Schnablegger H, Antonietti M (1999) J Colloid Interface Sci 212:197
85. Zheng J, Stevenson MS, Hikida RS, van Patten PG (2002) J Phys Chem B 106:1252
86. Zhou Y, Wang CY, Zhu YR, Chen ZY (1999) Chem Mater 11:2310
87. Mandal M, Ghosh SK, Kundu S, Esumi K, Pal T (2002) Langmuir 18:7792
88. Esumi K, Kameo A, Suzuki A, Torigoe K (2001) Colloids Surf A: Physiochem Eng Aspects 189:155
89. Esumi K, Torigoe K (2001) Prog Colloid Polym Sci 117:80
90. Bronstein LM, Sidorov SN, Gourkova AY, Valetsky PM, Hartmann J, Breulmann M, Cölfen H, Antonietti M (1998) Inorg Chim Acta 280:348
91. Esumi K, Suzuki A, Aihara N, Usui K, Torigoe K (1998) Langmuir 14:3157
92. Esumi K, Suzuki A, Yamahira A, Torigoe K (2000) Langmuir 16:2604
93. Wells M, Crooks RM (1996) J Am Chem Soc 118:3988
94. Haag R, Krämer M, Stumbé J-F, Krause S, Komp A, Prokhorva S (2002) Polymer Preprints 43:328
95. Aymonier C, Schlotterbeck U, Antonietti L, Zacharias P, Thomann R, Tiller JC, Mecking S (2002) Chem Commun, p 3018
96. Sunder A, Krämer M, Hanselmann R, Mülhaupt R, Frey H (1999) Angew Chem Int Ed 38:3552
97. Haag R, Stumbé J-F, Sunder A, Frey H, Hebel A (2000) Macromolecules 33:8158
98. Stiriba S-E, Kautz H, Frey H (2002) J Am Chem Soc 124:9698
99. Slagt MQ, Stiriba S-E, Gebbink RJMK, Kautz H, Frey H, Van Koten G (2002) Macromolecules 35:5734
100. Hirai H, Nakao Y, Toshima N (1979) J Macromol Sci Chem A13:727
101. Bönnemann H, Brijoux W, Brinkmann R, Fretzen R, Joussen T, Koeppler R, Korall B, Neiteler P, Richter J (1994) J Mol Catal 86:129
102. Antonietti M, Wenz E, Bronstein L, Seregina M (1995) Adv Mater 7:1000
103. Seregina MV, Bronstein LM, Platonova OA, Chernyshov DM, Valetsky PM, Hartmann J, Wenz E, Antonietti M (1997) Chem Mater 9:923
104. Mayer ABR, Mark JE (1997) J Polym Sci Polym Chem 35:3151
105. Mecking S, Schlotterbeck U, Thomann R, Soddemann M, Stieger M, Richtering W, Kautz H (2001) Polym Mater Sci Eng, Am Chem Soc 84:511

106. Haag R, Sunder A, Hebel A, Roller S (2002) J Comb Chem 4:112
107. Peinemann KV, Nunes SP (2001) In: Peinemann KV, Nunes SP (eds) Membrane Technology in the Chemical Industry. Wiley, Weinheim, p 25
108. Baars MWPL, Karlsson AJ, Sorokin V, De Waal BFM, Meijer EW (2000) Angew Chem Int Ed 39:4262
109. Boas U, Karlsson AJ, De Waal BFM, Meijer EW (2001) J Org Chem 66:2136
110. Johannsen M, Jorgensen KA (1998) Chem Rev 98:1689
111. Tzschucke CC, Markert C, Glatz H, Bannwarth W (2002) Angew Chem Int Ed 41:4500
112. Rieck H, Helmchen G (1995) Angew Chem Int Ed Engl 34:2687
113. Prétôt R, Lloyd-Jones GC, Pfaltz A (1998) Pure Appl Chem 70:1035
114. van Haaren RJ, Druijven CJM, van Strijdonck GPF, Oevering H, Reek JNH, Kamer PCJ, van Leeuwen PWNM (2000) J Chem Soc, Dalton Trans, p 1549
115. De Groot D, Eggeling EB, De Wilde JC, Kooijman H, Van Haaren RJ, Van der Made AW, Spek AL, Vogt D, Reek JNH, Kamer PCJ, Van Leeuwen PWNM (1999) Chem Commun, p 1623
116. Hovestad NJ, Eggeling EB, Heidbüchel HJ, Jastrzebski JTBH, Kragl U, Keim W, Vogt D, Van Koten G (1999) Angew Chem Int Ed 38:1655
117. Brinkmann N, Giebel D, Lohmer G, Reetz MT, Kragl U (1999) J Catal 183:163
118. Tzschucke CC, Bannwarth W (2004) Helv Chim Acta 87:2882
119. Garcia-Bernabé A, Krämer M, Olah B, Haag R (2004) Chem Eur J 10:2822
120. Garcia-Bernabé A, Tzschucke CC, Bannwarth W, Haag R (2005) Adv Synth Catal 347:1389

Dérien S, see Bruneau C (2006) *19*: 295–326

Derien S, see Dixneuf PH (2004) *11*: 1–44

Deubel D, Loschen C, Frenking G (2005) Organometallacycles as Intermediates in Oxygen-Transfer Reactions. Reality or Fiction? *12*: 109–144

Diéguez M, see Claver C (2006) *18*: 35–64

Dixneuf PH, Derien S, Monnier F (2004) Ruthenium-Catalyzed C–C Bond Formation. *11*: 1–44

Dixneuf PH, see Bruneau C (2006) *19*: 295–326

Dötz KH, Minatti A (2004) Chromium-Templated Benzannulation Reactions. *13*: 123–156

Dowdy EC, see Molander G (1999) *2*: 119–154

Dowdy ED, see Delmonte AJ (2004) *6*: 97–122

Doyle MP (2004) Metal Carbene Reactions from Dirhodium(II) Catalysts. *13*: 203–222

Drudis-Solé G, Ujaque G, Maseras F, Lledós A (2005) Enantioselectivity in the Dihydroxylation of Alkenes by Osmium Complexes. *12*: 79–107

Dyson PJ, see Allardyce CS (2006) *17*: 177–210

Eilbracht P, Schmidt AM (2006) Synthetic Applications of Tandem Reaction Sequences Involving Hydroformylation. *18*: 65–95

Eisen MS, see Lisovskii A (2005) *10*: 63–105

Fañanás FJ, see Barluenga (2004) *13*: 59–121

Fensterbank L, see Aubert C (2006) *19*: 259–294

Flórez J, see Barluenga (2004) *13*: 59–121

Fontecave M, Hamelin O, Ménage S (2005) Chiral-at-Metal Complexes as Asymmetric Catalysts. *15*: 271–288

Fontecilla-Camps JC, see Volbeda A (2006) *17*: 57–82

Fraile JM, García JI, Mayoral JA (2005) Non-covalent Immobilization of Catalysts Based on Chiral Diazaligands. *15*: 149–190

Frenking G, see Deubel D (2005) *12*: 109–144

Freund H-J, see Risse T (2005) *16*: 117–149

Fu GC, see Netherton M (2005) *14*: 85–108

Fujdala KL, Brutchey RL, Tilley TD (2005) Tailored Oxide Materials via Thermolytic Molecular Precursor (TMP) Methods. *16*: 69–115

Fürstner A (1998) Ruthenium-Catalyzed Metathesis Reactions in Organic Synthesis. *1*: 37–72

Gabriele B, Salerno G, Costa M (2006) Oxidative Carbonylations. *18*: 239–272

Gade LH, see Kassube JK (2006) *20*: 61–96

Gandon V, see Aubert C (2006) *19*: 259–294

García JI, see Fraile JM (2005) *15*: 149–190

Gates BC (2005) Oxide- and Zeolite-supported "Molecular" Metal Clusters: Synthesis, Structure, Bonding, and Catalytic Properties. *16*: 211–231

Gibson SE (née Thomas), Keen SP (1998) Cross-Metathesis. *1*: 155–181

Gilbertson JD, see Chandler BD (2006) *20*: 97–120

Gisdakis P, see Rösch N (1999) *4*: 109–163

Görling A, see Rösch N (1999) *4*: 109–163

Goldfuss B (2003) Enantioselective Addition of Organolithiums to C=O Groups and Ethers. *5*: 12–36

Gossage RA, van Koten G (1999) A General Survey and Recent Advances in the Activation of Unreactive Bonds by Metal Complexes. *3*: 1–8

Gotov B, see Schmalz HG (2004) 7: 157–180
Gras E, see Hodgson DM (2003) 5: 217–250
Grepioni F, see Braga D (1999) 4: 47–68
Gröger H, see Shibasaki M (1999) 2: 199–232
Groppo E, see Zecchina A (2005) 16: 1–35
Grushin VV, Alper H (1999) Activation of Otherwise Unreactive C–Cl Bonds. 3: 193–225
Guitian E, Perez D, Pena D (2005) Palladium-Catalyzed Cycloaddition Reactions of Arynes.
      14: 109–146

Haag R, see Hajji C (2006) 20: 149–176
Hajji C, Haag R (2006) Hyperbranched Polymers as Platforms for Catalysts. 20: 149–176
Hamelin O, see Fontecave M (2005) 15: 271–288
Harman D (2004) Dearomatization of Arenes by Dihapto-Coordination. 7: 95–128
Hatano M, see Mikami M (2005) 14: 279–322
Haynes A (2006) Acetic Acid Synthesis by Catalytic Carbonylation of Methanol. 18: 179–205
He Y, see Nicolaou KC (1998) 1: 73–104
Hegedus LS (2004) Photo-Induced Reactions of Metal Carbenes in organic Synthesis. 13:
      157–201
Hermanns J, see Schmidt B (2004) 13: 223–267
Hidai M, Mizobe Y (1999) Activation of the N–N Triple Bond in Molecular Nitrogen: Toward
      its Chemical Transformation into Organo-Nitrogen Compounds. 3: 227–241
Hirao T, see Moriuchi T (2006) 17: 143–175
Hodgson DM, Stent MAH (2003) Overview of Organolithium-Ligand Combinations and
      Lithium Amides for Enantioselective Processes. 5: 1–20
Hodgson DM, Tomooka K, Gras E (2003) Enantioselective Synthesis by Lithiation Adjacent
      to Oxygen and Subsequent Rearrangement. 5: 217–250
Hoppe D, Marr F, Brüggemann M (2003) Enantioselective Synthesis by Lithiation Adjacent
      to Oxygen and Electrophile Incorporation. 5: 61–138
Hou Z, Wakatsuki Y (1999) Reactions of Ketones with Low-Valent Lanthanides: Isolation
      and Reactivity of Lanthanide Ketyl and Ketone Dianion Complexes. 2: 233–253
Hoveyda AH (1998) Catalytic Ring-Closing Metathesis and the Development of Enantio-
      selective Processes. 1: 105–132
Huang M, see Wu GG (2004) 6: 1–36
Hughes DL (2004) Applications of Organotitanium Reagents. 6: 37–62

Iguchi M, Yamada K, Tomioka K (2003) Enantioselective Conjugate Addition and 1,2-Addition
      to C=N of Organolithium Reagents. 5: 37–60
Ito Y, see Murakami M (1999) 3: 97–130
Ito Y, see Suginome M (1999) 3: 131–159
Itoh K, Yamamoto Y (2004) Ruthenium Catalyzed Synthesis of Heterocyclic Compounds.
      11: 249–276

Jacobi von Wangelin A, Neumann H, Beller M (2006) Carbonylations of Aldehydes. 18:
      207–221
Jacobsen EN, see Larrow JF (2004) 6: 123–152
Johnson TA, see Break P (2003) 5: 139–176
Jones WD (1999) Activation of C–H Bonds: Stoichiometric Reactions. 3: 9–46

Kagan H, Namy JL (1999) Influence of Solvents or Additives on the Organic Chemistry
      Mediated by Diiodosamarium. 2: 155–198

Negishi E, Wang G, Zhu G (2006) Palladium-Catalyzed Cyclization via Carbopalladation and Acylpalladation. *19*: 1–48

Netherton M, Fu GC (2005) Palladium-catalyzed Cross-Coupling Reactions of Unactivated Alkyl Electrophiles with Organometallic Compounds. *14*: 85–108

Neumann H, see Jacobi von Wangelin A (2006) *18*: 207–221

Nicolaou KC, King NP, He Y (1998) Ring-Closing Metathesis in the Synthesis of Epothilones and Polyether Natural Products. *1*: 73–104

Nishiyama H (2004) Cyclopropanation with Ruthenium Catalysts. *11*: 81–92

Noels A, Demonceau A, Delaude L (2004) Ruthenium Promoted Catalysed Radical Processes toward Fine Chemistry. *11*: 155–171

Nolan SP, Viciu MS (2005) The Use of N-Heterocyclic Carbenes as Ligands in Palladium Mediated Catalysis. *14*: 241–278

Normant JF (2003) Enantioselective Carbolithiations. *5*: 287–310

Norton JR, see Cummings SA (2005) *10*: 1–39

Nozaki K, see Nakano K (2006) *18*: 223–238

Oberholzer MR, see Bien J (2004) *6*: 263–284

Obst D, see Wiese K-D (2006) *18*: 1–33

Öhler E, see Mulzer J (2004) *13*: 269–366

Pàmies O, see Claver C (2006) *18*: 35–64

Pape A, see Kündig EP (2004) *7*: 71–94

Patil NT, Yamamoto Y (2006) Palladium Catalyzed Cascade Reactions Involving π-Allyl Palladium Chemistry. *19*: 91–114

Pawlow JH, see Tindall D, Wagener KB (1998) *1*: 183–198

Pena D, see Guitian E (2005) *14*: 109–146

Perez D, see Guitian E (2005) *14*: 109–146

Pérez-Castells J (2006) Cascade Reactions Involving Pauson–Khand and Related Processes. *19*: 207–258

Prashad M (2004) Palladium-Catalyzed Heck Arylations in the Synthesis of Active Pharmaceutical Ingredients. *6*: 181–204

Prestipino C, see Zecchina A (2005) *16*: 1–35

Pretraszuk C, see Marciniec B (2004) *11*: 197–248

Reek JNH, see Ribaudo F (2006) *20*: 39–59

Riant O, see Chuzel O (2005) *15*: 59–92

Ribaudo F, van Leeuwen PWNM, Reek JNH (2006) Supramolecular Dendritic Catalysis: Noncovalent Catalyst Anchoring to Functionalized Dendrimers. *20*: 39–59

Richmond TG (1999) Metal Reagents for Activation and Functionalization of Carbon–Fluorine Bonds. *3*: 243–269

Rigby J, Kondratenkov M (2004) Arene Complexes as Catalysts. *7*: 181–204

Risse T, Freund H-J (2005) Spectroscopic Characterization of Organometallic Centers on Insulator Single Crystal Surfaces: From Metal Carbonyls to Ziegler–Natta Catalysts. *16*: 117–149

Rodríguez F, see Barluenga (2004) *13*: 59–121

Roland S, Mangeney P (2005) Chiral Diaminocarbene Complexes, Synthesis and Application in Asymmetric Catalysis. *15*: 191–229

Rösch N (1999) A Critical Assessment of Density Functional Theory with Regard to Applications in Organometallic Chemistry. *4*: 109–163

Roucoux A (2005) Stabilized Noble Metal Nanoparticles: An Unavoidable Family of Catalysts for Arene Derivative Hydrogenation. *16*: 261–279

Ruiz J, see Astruc D (2006) *20*: 121–148

Sakaki S (2005) Theoretical Studies of C–H s-Bond Activation and Related by Transition-Metal Complexes. *12*: 31–78

Salerno G, see Gabriele B (2006) *18*: 239–272

Satoh T, see Miura M (2005) *14*: 1–20

Satoh T, see Miura M (2005) *14*: 55–84

Savoia D (2005) Progress in the Asymmetric Synthesis of 1,2-Diamines from Azomethine Compounds. *15*: 1–58

Schmalz HG, Gotov B, Böttcher A (2004) Natural Product Synthesis. *7*: 157–180

Schmidt AM, see Eilbracht P (2006) *18*: 65–95

Schmidt B, Hermanns J (2004) Olefin Metathesis Directed to Organic Synthesis: Principles and Applications. *13*: 223–267

Schrock RR (1998) Olefin Metathesis by Well-Defined Complexes of Molybdenum and Tungsten. *1*: 1–36

Schulz E (2005) Use of *N,N*-Coordinating Ligands in Catalytic Asymmetric C–C Bond Formations: Example of Cyclopropanation, Diels–Alder Reaction, Nucleophilic Allylic Substitution. *15*: 93–148

Semmelhack MF, Chlenov A (2004) (Arene)Cr(Co)$_3$ Complexes: Arene Lithiation/Reaction with Electrophiles. *7*: 21–42

Semmelhack MF, Chlenov A (2004) (Arene)Cr(Co)$_3$ Complexes: Aromatic Nucleophilic Substitution. *7*: 43–70

Sen A (1999) Catalytic Activation of Methane and Ethane by Metal Compounds. *3*: 81–95

Severin K (2006) Organometallic Receptors for Biologically Interesting Molecules. *17*: 123–142

Sheldon RA, see Arends IWCE (2004) *11*: 277–320

Shibasaki M, Gröger H (1999) Chiral Heterobimetallic Lanthanoid Complexes: Highly Efficient Multifunctional Catalysts for the Asymmetric Formation of C–C, C–O and C–P Bonds. *2*: 199–232

Simonneaux G, Le Maux P (2006) Carbene Complexes of Heme Proteins and Iron Porphyrin Models. *17*: 83–122

Staemmler V (2005) The Cluster Approach for the Adsorption of Small Molecules on Oxide Surfaces. *12*: 219–256

Stent MAH, see Hodgson DM (2003) *5*: 1–20

Strassner T (2004) Electronic Structure and Reactivity of Metal Carbenes. *13*: 1–20

Strong LE, see Kiessling LL (1998) *1*: 199–231

Strübing D, Beller M (2006) The Pauson–Khand Reaction. *18*: 165–178

Suginome M, Ito Y (1999) Activation of Si–Si Bonds by Transition-Metal Complexes. *3*: 131–159

Sumi K, Kumobayashi H (2004) Rhodium/Ruthenium Applications. *6*: 63–96

Suzuki N (2005) Stereospecific Olefin Polymerization Catalyzed by Metallocene Complexes. *8*: 177–215

Szymoniak J, Bertus P (2005) Zirconocene Complexes as New Reagents for the Synthesis of Cyclopropanes. *10*: 107–132

Takahashi T, Kanno K (2005) Carbon–Carbon Bond Cleavage Reaction Using Metallocenes. *8*: 217–236

Tan Z, see Negishi E (2005) *8*: 139–176

Tilley TD, see Fujdala KL (2005) *16*: 69–115

Tindall D, Pawlow JH, Wagener KB (1998) Recent Advances in ADMET Chemistry. *1*: 183–198

Tobisch S (2005) Co-Oligomerization of 1,3-Butadiene and Ethylene Promoted by Zerovalent 'Bare' Nickel Complexes. *12*: 187–218

Tomioka K, see Iguchi M (2003) *5*: 37–60

Tomooka K, see Hodgson DM (2003) *5*: 217–250

Toniolo L, see Cavinato G (2006) *18*: 125–164

Toru T, Nakamura S (2003) Enantioselective Synthesis by Lithiation Adjacent to Sulfur, Selenium or Phosphorus, or without an Adjacent Activating Heteroatom. *5*: 177–216

Tunge JA, see Cummings SA (2005) *10*: 1–39

Uemura M (2004) (Arene)Cr(Co)₃ Complexes: Cyclization, Cycloaddition and Cross Coupling Reactions. *7*: 129–156

Ujaque G, see Drudis-Solé G (2005) *12*: 79–107

Urrutigoïty M, see Kalck P (2006) *18*: 97–123

Vavasori A, see Cavinato G (2006) *18*: 125–164

Viciu MS, see Nolan SP (2005) *14*: 241–278

Volbeda A, Fontecilla-Camps JC (2006) Catalytic Nickel–Iron–Sulfur Clusters: From Minerals to Enzymes. *17*: 57–82

Wagener KB, see Tindall D, Pawlow JH (1998) *1*: 183–198

Wakatsuki Y, see Hou Z (1999) *2*: 233–253

Wang M, see Li CJ (2004) *11*: 321–336

Wang G, see Negishi E (2006) *19*: 1–48

Watson DJ, see Delmonte AJ (2004) *6*: 97–122

Wiese K-D, Obst D (2006) Hydroformylation. *18*: 1–33

Wipf P, Kendall C (2005) Hydrozirconation and Its Applications. *8*: 1–25

Wu GG, Huang M (2004) Organolithium in Asymmetric Process. *6*: 1–36

Wu YT, de Meijere A (2004) Versatile Chemistry Arising from Unsaturated Metal Carbenes. *13*: 21–58

Xi Z, Li Z (2005) Construction of Carbocycles via Zirconacycles and Titanacycles. *8*: 27–56

Yamada K, see Iguchi M (2003) *5*: 37–60

Yamamoto A, see Lin Y-S (1999) *3*: 161–192

Yamamoto Y, Nakamura I (2005) Nucleophilic Attack by Palladium Species. *14*: 211–240

Yamamoto Y, see Itoh K (2004) *11*: 249–276

Yamamoto Y, see Patil NT (2006) *19*: 91–114

Yasuda H (1999) Organo Rare Earth Metal Catalysis for the Living Polymerizations of Polar and Nonpolar Monomers. *2*: 255–283

Yasuda N, see King AO (2004) *6*: 205–246

Zecchina A, Groppo E, Damin A, Prestipino C (2005) Anatomy of Catalytic Centers in Phillips Ethylene Polymerization Catalyst. *16*: 1–35

von Zezschwitz P, de Meijere A (2006) Domino Heck-Pericyclic Reactions. *19*: 49–90

Zhu G, see Negishi E (2006) *19*: 1–48

Ziegler T, see Michalak A (2005) *12*: 145–186

Subject Index

Printing: Krips bv, Meppel
Binding: Stürtz, Würzburg